Limiting the Magnitude of Future Climate Change

America's Climate Choices
Panel on Limiting the Magnitude of Climate Change

Board on Atmospheric Sciences and Climate

Division on Earth and Life Studies

NATIONAL RESEARCH COUNCIL
OF THE NATIONAL ACADEMIES

THE NATIONAL ACADEMIES PRESS
Washington, D.C.
www.nap.edu

THE NATIONAL ACADEMIES PRESS · 500 Fifth Street, N.W. · Washington, DC 20001

NOTICE: The project that is the subject of this report was approved by the Governing Board of the National Research Council, whose members are drawn from the councils of the National Academy of Sciences, the National Academy of Engineering, and the Institute of Medicine. The members of the committee responsible for the report were chosen for their special competences and with regard for appropriate balance.

This study was supported by the National Oceanic and Atmospheric Administration under contract number DG133R08CQ0062, TO# 4. Any opinions, findings, conclusions, or recommendations expressed in this material are those of the author(s) and do not necessarily reflect the views of the sponsoring agency or any of its subagencies.

International Standard Book Number-13: 978-0-309-14597-8 (Book)
International Standard Book Number-10: 0-309-14597-X (Book)
International Standard Book Number-13: 978-0-309-14598-5 (PDF)
International Standard Book Number-10: 0-309-14598-8 (PDF)
Library of Congress Control Number: 2010940141

Additional copies of this report are available from the National Academies Press, 500 Fifth Street, N.W., Lockbox 285, Washington, DC 20055; (800) 624-6242 or (202) 334-3313 (in the Washington metropolitan area); Internet, http://www.nap.edu

Cover images:

Middle left: courtesy of www.public-domain-image.com

Far right: courtesy of Department of Housing

THE NATIONAL ACADEMIES
Advisers to the Nation on Science, Engineering, and Medicine

The **National Academy of Sciences** is a private, nonprofit, self-perpetuating society of distinguished scholars engaged in scientific and engineering research, dedicated to the furtherance of science and technology and to their use for the general welfare. Upon the authority of the charter granted to it by the Congress in 1863, the Academy has a mandate that requires it to advise the federal government on scientific and technical matters. Dr. Ralph J. Cicerone is president of the National Academy of Sciences.

The **National Academy of Engineering** was established in 1964, under the charter of the National Academy of Sciences, as a parallel organization of outstanding engineers. It is autonomous in its administration and in the selection of its members, sharing with the National Academy of Sciences the responsibility for advising the federal government. The National Academy of Engineering also sponsors engineering programs aimed at meeting national needs, encourages education and research, and recognizes the superior achievements of engineers. Dr. Charles M. Vest is president of the National Academy of Engineering.

The **Institute of Medicine** was established in 1970 by the National Academy of Sciences to secure the services of eminent members of appropriate professions in the examination of policy matters pertaining to the health of the public. The Institute acts under the responsibility given to the National Academy of Sciences by its congressional charter to be an adviser to the federal government and, upon its own initiative, to identify issues of medical care, research, and education. Dr. Harvey V. Fineberg is president of the Institute of Medicine.

The **National Research Council** was organized by the National Academy of Sciences in 1916 to associate the broad community of science and technology with the Academy's purposes of furthering knowledge and advising the federal government. Functioning in accordance with general policies determined by the Academy, the Council has become the principal operating agency of both the National Academy of Sciences and the National Academy of Engineering in providing services to the government, the public, and the scientific and engineering communities. The Council is administered jointly by both Academies and the Institute of Medicine. Dr. Ralph J. Cicerone and Dr. Charles M. Vest are chair and vice chair, respectively, of the National Research Council.

www.national-academies.org

Foreword: About America's Climate Choices

Convened by the National Research Council in response to a request from Congress (P.L. 110-161), *America's Climate Choices* is a suite of five coordinated activities designed to study the serious and sweeping issues associated with global climate change, including the science and technology challenges involved, and to provide advice on the most effective steps and most promising strategies that can be taken to respond.

The Committee on America's Climate Choices is responsible for providing overall direction, coordination, and integration of the *America's Climate Choices* suite of activities and ensuring that these activities provide well-supported, action-oriented, and useful advice to the nation. The committee convened a Summit on America's Climate Choices on March 30–31, 2009, to help frame the study and provide an opportunity for high-level input on key issues. The committee is also charged with writing a final report that builds on four panel reports and other sources to answer the following four overarching questions:

- What short-term actions can be taken to respond effectively to climate change?
- What promising long-term strategies, investments, and opportunities could be pursued to respond to climate change?
- What are the major scientific and technological advances needed to better understand and respond to climate change?
- What are the major impediments (e.g., practical, institutional, economic, ethical, intergenerational, etc.) to responding effectively to climate change, and what can be done to overcome these impediments?

The Panel on Limiting the Magnitude of Future Climate Change was charged to describe, analyze, and assess strategies for reducing the net future human influence on climate. This report focuses on actions to reduce domestic greenhouse gas emissions and other human drivers of climate change, such as changes in land use, but also considers the international dimensions of climate stabilization.

The Panel on Adapting to the Impacts of Climate Change was charged to describe, analyze, and assess actions and strategies to reduce vulnerability, increase adaptive capacity, improve resiliency, and promote successful adaptation to climate change in different regions, sectors, systems, and populations. The panel's report draws on a wide range of sources and case studies to identify lessons learned from past experiences, promising current approaches, and potential new directions.

The Panel on Advancing the Science of Climate Change was charged to provide a concise overview of past, present, and future climate change, including its causes and its impacts, and to recommend steps to advance our current understanding, including new observations, research programs, next-generation models, and the physical and human assets needed to support these and other activities. The panel's report focuses on the scientific advances needed both to improve our understanding of the integrated human-climate system and to devise more effective responses to climate change.

The Panel on Informing Effective Decisions and Actions Related to Climate Change was charged to describe and assess different activities, products, strategies, and tools for informing decision makers about climate change and helping them plan and ex-ecute effective, integrated responses. The panel's report describes the different types of climate change-related decisions and actions being taken at various levels and in different sectors and regions; it develops a framework, tools, and practical advice for ensuring that the best available technical knowledge about climate change is used to inform these decisions and actions.

America's Climate Choices builds on an extensive foundation of previous and ongoing work, including National Research Council reports, assessments from other national and international organizations, the current scientific literature, climate action plans by various entities, and other sources. More than a dozen boards and standing com-mittees of the National Research Council were involved in developing the study, and many additional groups and individuals provided additional input during the study process. Outside viewpoints were also obtained via public events and workshops (including the Summit), invited presentations at committee and panel meetings, and comments received through the study website, *http://americasclimatechoices.org*.

Collectively, the *America's Climate Choices* suite of activities involves more than 90 volunteers from a range of communities including academia, various levels of govern-ment, business and industry, other nongovernmental organizations, and the interna-tional community. Responsibility for the final content of each report rests solely with the authoring panel and the National Research Council. However, the development of each report included input from and interactions with members of all five study groups; the membership of each group is listed in Appendix A.

Preface

Tackling climate change promises to be one of the most significant social and technological challenges of the 21st century. Since the industrial revolution, the atmosphere has been one of the world's principal waste repositories because it has offered an easy and inexpensive means of managing unwanted by-products. It is currently absorbing a net gain of two parts per million of CO_2 per year as the result of global emissions, and the world's leading scientists believe that this change in atmospheric composition is changing the global climate.

This report focuses on actions available to the United States to reduce greenhouse gas (GHG) emissions. The goal of actually limiting[1] global climate change requires international cooperation, since most of this century's emissions will come from developing countries, with U.S. emissions representing a shrinking portion of the total. Thus, our national strategy must promote domestic actions while at the same time influencing the rest of the world to control their emissions.

The United States has successfully reduced emissions of several key atmospheric pollutants—including SO_2, NO_X, and particulates—through the Clean Air Act. The creation of a market for SO_2 allowances, in conjunction with performance standards and a cap on emissions, provided strong incentives for entrepreneurs to develop lower-cost SO_2 abatement technologies and approaches, and is one of the past century's greatest environmental policy successes. Emissions of most GHGs, however, remain largely unregulated and continue to be discharged without penalty, through smokestacks, tailpipes, and chimneys, and by the destruction of forests. With no price on carbon, or regulatory pressure, there exist few incentives to mitigate emissions. Thus, we continue to "lock in" incumbent technologies and systems that are typically carbon-intensive. Changing these practices will require scientific and engineering genius to create new energy systems that avoid emitting all but a small fraction of today's GHGs while simultaneously powering global economic growth. Success will also necessitate institutional, economic, social, and policy innovations to foster the widespread and rapid deployment of transformational technologies.

---

[1] The term "limiting" climate change rather than "mitigation" of climate change was deliberately chosen, because in some circles, mitigation often refers to mitigating the impacts of climate change, that is, adaptation (the focus of another *America's Climate Choices* panel report). Our focus is on limiting the main drivers of climate change (i.e., greenhouse gas emissions), with the expectation that this will contribute to limiting climate change itself.

In this study, the panel was charged with describing, analyzing, and assessing strategies for reducing the future human influence on climate (see full Statement of Task in Appendix B). We considered both existing and emerging technologies, as well as existing and innovative new policies. Technologies and policies were assessed according to their scale of impact, cost, feasibility, and other critical factors, with the assistance of a set of guiding principles. Based on these factors and principles, the panel was able to recommend a short list of options that appear to be most important for significantly reducing GHG emissions.

There are numerous important issues closely related to the topic of limiting climate change that are not addressed here. This report does not, for instance,

- describe the scientific evidence for why climate change is real and being driven largely by human influences and why this poses a serious threat to humans and ecosystems;
- identify the impacts that may result from not taking sufficient action to limit climate change and the vulnerability of different populations and regions to those impacts;
- analyze the economic impacts of acting versus not acting to limit climate change (i.e., cost-benefit analyses);
- discuss "solar radiation management" geoengineering strategies;
- explore (in any considerable depth) the scientific research needed for improving our understanding of climate change and the specific types of technological research and development needed for reducing emissions; or
- examine strategies for improving education and communication about climate change with the general public and the media.

Many of these issues are addressed in the other *America's Climate Choices* panel reports (*Advancing the Science of Climate Change, Adapting to the Impacts of Climate Change, Informing an Effective Response to Climate Change*), and/or will be addressed in the final report of the ACC main committee.

This study began at a time when the United States and countries around the world were actively debating options for addressing global climate change. It is particularly timely, therefore, that the National Research Council has taken on this task; in doing so, we were fortunate to engage a panel of experts with a diversity of backgrounds—including, for instance, physical scientists, social scientists, economists, engineers, community organizers, lawyers, and executives of nongovernmental organizations. This broad-based group of experts proved capable of resolving many opposing viewpoints that at first blush might have seemed irreconcilable. Their active involvement and commitment to producing a useful report is greatly appreciated.

The panel approached its task by conducting its own review of the literature and by supplementing the panel members' expertise with informational briefings on key topics from outside authorities. In particular, we wish to thank Jonathan Black, Senate Energy and Natural Resources Committee; Rachel Cleetus and Steve Clemmer, Union of Concerned Scientists; Ana Unruh Cohen, House Select Committee on Energy Independence and Global Warming; Robert Marlay, Climate Change Technology Program, U.S. Department of Energy; W. David Montgomery, CRA International; Bill Parton, Colorado State University; Robert Pollin, University of Massachusetts, Amherst; and Michael Ryan, USDA Forest Service.

We are particularly grateful for the assistance provided by Laurie Geller, who managed this panel study for the National Research Council. Her unflagging persistence, upbeat attitude, regular communications, and writing and editing assistance helped keep the panel on schedule. She received considerable assistance from Tom Menzies, Alan Crane, and Paul Stern, who were important sounding boards as the panel's ideas were being formulated. We also thank Shelly Freeland, who managed the logistics of our meetings, and Katie Weller, who supported the preparation of the final manuscript.

Bob Fri (*Chair*) and Marilyn Brown (*Vice Chair*)
America's Climate Choices:
Panel on Limiting the Magnitude of Future Climate Change

Acknowledgments

This report has been reviewed in draft form by individuals chosen for their diverse perspectives and technical expertise, in accordance with procedures approved by the National Research Council's (NRC's) Report Review Committee. The purpose of this independent review is to provide candid and critical comments that will assist the institution in making its published report as sound as possible and to ensure that the report meets institutional standards for objectivity, evidence, and responsiveness to the study charge. The review comments and draft manuscript remain confidential to protect the integrity of the deliberative process. We wish to thank the following individuals for their review of this report:

PAUL DECOTIS, Long Island Power Authority
PETER FRUMHOFF, Union of Concerned Scientists
ARNULF GRUBLER, International Institute for Applied Systems Analysis
HENRY JACOBY, Massachusetts Institute of Technology
ROGER KASPERSON, Clark University
FRANZ LITZ, World Resources Institute
WILLIAM NORDHAUS, Yale University
MICHAEL OPPENHEIMER, Princeton University
ROBERT POLLIN, University of Massachusetts, Amherst
MAXINE SAVITZ, Honeywell, Inc. (retired)
RICHARD SCHMALENSEE, Massachusetts Institute of Technology
ROBERT SOCOLOW, Princeton University
BJORN STIGSON, World Business Council for Sustainable Development
MICHAEL VANDENBERGH, Vanderbilt University
DAVID VICTOR, Stanford University

Although the reviewers listed above have provided many constructive comments and suggestions, they were not asked to endorse the conclusions or recommendations nor did they see the final draft of the report before its release. The review of this report was overseen by **Robert Frosch** (Harvard University) and **Tom Graedel** (Yale University). Appointed by the NRC, they were responsible for making certain that an independent examination of this report was carried out in accordance with institutional procedures and that all review comments were carefully considered. Responsibility for the final content of this report rests entirely with the authoring panel and the institution.

Institutional oversight for this project was provided by:

SHELLY FREELAND, Senior Program Assistant
AMANDA PURCELL, Senior Program Assistant
JANEISE STURDIVANT, Program Assistant
RICARDO PAYNE, Program Assistant
SHUBHA BANSKOTA, Financial Associate

Contents

SUMMARY   1

1   INTRODUCTION   15
      Context and Purpose of This Report, 15
      Principles to Guide Climate Change Limiting Policy and Strategy, 17
      Organization of the Report, 18

2   GOALS FOR LIMITING FUTURE CLIMATE CHANGE   21
      Reference U.S. and Global Emissions, 21
      Setting Climate Change Limiting Goals, 30
      Global Emission Targets, 32
      U.S. Emission Targets, 36
      Implications of U.S. Emission Goals, 42
      Key Conclusions and Recommendations, 48

3   OPPORTUNITIES FOR LIMITING FUTURE CLIMATE CHANGE   51
      Opportunities for Limiting GHG Emissions, 51
      The Case for Urgency, 81
      The Larger Context for Technology, 87
      Key Conclusions and Recommendations, 88

4   CRAFTING A PORTFOLIO OF CLIMATE CHANGE LIMITING POLICIES   91
      Pricing Strategy Design Features, 92
      Comparing Taxes with Cap and Trade, 99
      Complementary Options for the Policy Portfolio, 108
      Integrating the Policy Options, 126
      Key Conclusions and Recommendations, 133

5   FOSTERING TECHNOLOGICAL INNOVATIONS   137
      The Role of Technological Innovation, 137
      The Process of Technological Change, 139
      Resources Currently Available for Technology Innovation, 143
      Assessment of Current U.S. Innovation Policies, 157
      Key Conclusions and Recommendations, 163

CONTENTS

6 INTERACTION WITH OTHER MAJOR POLICY CONCERNS   165
   Linkages with Energy and Environmental Policy Issues, 165
   Key Conclusions and Recommendations, 172
   Equity and Employment Impacts, 173
   Key Conclusions and Recommendations, 190

7 MULTILEVEL RESPONSE STRATEGIES   193
   International Strategies, 193
   Key Conclusions and Recommendations, 205
   Balancing Federal with State and Local Action, 206
   Key Conclusions and Recommendations, 213

8 POLICY DURABILITY AND ADAPTABILITY   215
   Policy Stability, Durability, and Enforcement, 215
   Generating Timely Information for Adaptive Management, 219
   Key Conclusions and Recommendations, 222

REFERENCES   225

APPENDIXES

A America's Climate Choices: Membership Lists   239
B Panel on Limiting the Magnitude of Future Climate Change: Statement of Task 243
C Panel on Limiting the Magnitude of Future Climate Change:
   Biographical Sketches   245
D Acronyms, Energy Units, and Chemical Formulas   255

Summary

I n the legislation calling for an assessment of America's climate choices, Congress directed the National Research Council (NRC) to "investigate and study the serious and sweeping issues relating to global climate change and make recommendations regarding the steps that must be taken and what strategies must be adopted in response to global climate change." As part of the response to this request, the America's Climate Choices Panel on Limiting the Magnitude of Future Climate Change was charged to "describe, analyze, and assess strategies for reducing the net future human influence on climate, including both technology and policy options, focusing on actions to reduce domestic greenhouse gas (GHG) emissions and other human drivers of climate change, but also considering the international dimensions of climate stabilization" (see Appendix B for the full statement of task).

Our panel responded to this charge by evaluating the choices available for the United States to contribute to the global effort of limiting future climate change. More specifically, the panel focused on strategies to reduce concentrations of GHGs in the atmosphere, including strategies that are technically and economically feasible in the near term, as well as strategies that could potentially play an important role in the longer term.

Because policy that limits climate change is highly complex and involves a wide array of political and ethical considerations, scientific analysis does not always point to unequivocal answers. We offer specific recommendations in cases where research clearly shows that certain strategies and policy options are particularly effective; but in other cases, we simply discuss the range of possible choices available to decision makers. On the broadest level, we conclude that the United States needs the following:

- *Prompt and sustained strategies to reduce GHG emissions.* There is a need for policy responses to promote the technological and behavioral changes necessary for making substantial near-term GHG emission reductions. There is also a need to aggressively promote research, development, and deployment of new technologies, both to enhance our chances of making the needed emissions reductions and to reduce the costs of doing so.
- An inclusive *national framework for instituting response strategies and policies.* National policies for limiting climate change are implemented through the actions of private industry, governments at all levels, and millions of households

and individuals. The essential role of the federal government is to put in place an overarching, national policy framework designed to ensure that all of these actors are furthering the shared national goal of emissions reductions. In addition, a national policy framework that both generates and is underpinned by international cooperation is crucial if the risks of global climate change are to be substantially curtailed.

- *Adaptable means for managing policy responses.* It is inevitable that policies put in place now will need to be modified in the future as new scientific information emerges, providing new insights and understanding of the climate problem. Even well-conceived policies may experience unanticipated difficulties, while others may yield unexpectedly high levels of success. Moreover, the degree, rate, and direction of technological innovation will alter the array of response options available and the costs of emissions abatement. Quickly and nimbly responding to new scientific information, the state of technology, and evidence of policy effectiveness will be essential to successfully managing climate risks over the course of decades.

While recognizing that there is ongoing debate about the goals for international efforts to limit climate change, for this analysis we have focused on a range of global atmospheric GHG concentrations between 450 and 550 parts per million (ppm) CO_2-equivalent (eq), a range that has been extensively analyzed by the scientific and economic communities and is a focus of international climate policy discussions. In evaluating U.S. climate policy choices, it useful to set goals that are consistent with those in widespread international use, both for policy development and for making quantitative assessments of alternative strategies.

Global temperature and GHG concentration targets are needed to help guide long-term global action. Domestic policy, however, requires goals that are more directly linked to outcomes that can be measured and affected by domestic action. The panel thus recommends that the U.S. policy goal be stated as a quantitative limit on domestic GHG emissions over a specified time period—in other words, a GHG emissions budget.

The panel does not attempt to recommend a specific budget number, because there are many political and ethical judgments involved in determining an "appropriate" U.S. share of global emissions. As a basis for developing and assessing domestic strategies, however, the panel used recent integrated assessment modeling studies[1] to suggest

---

[1] Specifically, we drew upon the Energy Modeling Forum 22 studies (*http://emf.stanford.edu/research/emf22/*). See Box 2.2 for details about this study and the reasons it was deemed particularly useful for our purposes.

that a reasonable "representative" range for a domestic emissions budget would be 170 to 200 gigatons (Gt) of CO_2-eq for the period 2012 through 2050. This corresponds roughly to a reduction of emissions from 1990 levels by 80 to 50 percent, respectively. We note that this budget range is based on "global least cost" economic efficiency criteria for allocating global emissions among countries. Using other criteria, different budget numbers could be suggested. (For instance, some argue that, based on global "fairness" concerns, a more aggressive U.S. emission-reduction effort is warranted.)

As illustrated in Figure S.1, meeting an emissions budget in the range suggested above, especially the more stringent budget of 170 Gt CO_2-eq, will require a major departure from business-as-usual emission trends (in which U.S. emissions have been rising at a rate of ~1 percent per year for the past three decades). The main drivers of GHG emissions are population growth and economic activity, coupled with energy use per capita and per unit of economic output ("energy intensity"). Although the energy intensity of the U.S. economy has been improving for the past two decades, total emis-

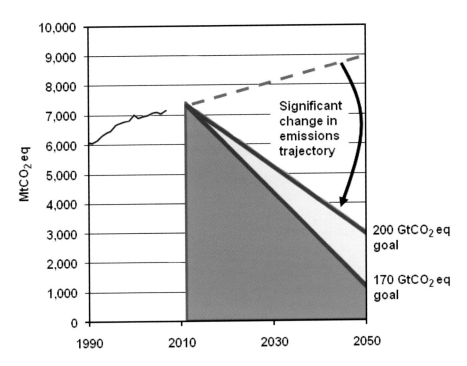

FIGURE S.1 Illustration of the representative U.S. cumulative GHG emissions budget targets: 170 and 200 Gt CO_2-eq (for Kyoto gases) (Gt, gigatons, or billion tons; Mt, megatons, or million tons). The exact value of the reference budget is uncertain, but nonetheless illustrates a clear need for a major departure from business as usual.

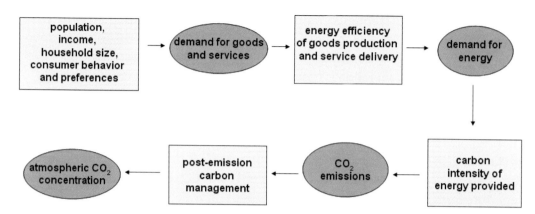

FIGURE S.2 The chain of factors that determine how much CO_2 accumulates in the atmosphere. The blue boxes represent factors that can potentially be influenced to affect the outcomes in the purple circles.

sions will continue to rise without a significant change from business as usual. Our analyses thus indicate that, without prompt action, the current rate of GHG emissions from the energy sector would consume the domestic emissions budget well before 2050.[2]

More than 80 percent of U.S. GHG emissions are in the form of CO_2 from combustion of fossil fuels. As illustrated in Figure S.2, there is a range of different opportunities to reduce emissions from the energy system, including reducing demand for goods and services requiring energy, improving the efficiency with which the energy is used to provide these goods and services, and reducing the carbon intensity of this energy supply (e.g., replacing fossil fuels with renewables or nuclear power, or employing carbon capture and storage).

To evaluate the magnitude and feasibility of needed changes in all of these areas, we examined the results of a recent NRC study, *America's Energy Future*, which estimated the technical potential for aggressive near-term (i.e., for 2020 and 2035) deployment of key technologies for energy efficiency and low-carbon energy production. We compared this to estimates of the technology deployment levels that might be needed to meet the representative emissions budget. This analysis suggests that limiting domestic GHG emissions to 170 Gt CO_2-eq by 2050 by relying only on these near-term opportunities may be technically possible but will be very difficult. Meeting the 200-Gt CO_2-eq goal will be somewhat less difficult but also very demanding. In either case, however, falling short of the full technical potential for technology deployment is

---

[2]For reference, U.S. GHG emissions for 2008 (latest year available) were approximately 7 Gt CO_2-eq.

likely, as it would require overcoming many existing barriers (e.g., social resistance and institutional and regulatory concerns).

Some important opportunities exist to control non-CO_2 GHGs (such as methane, nitrous oxide, and the long-lived fluorinated gases) and to enhance biological uptake of CO_2 through afforestation and tillage change on suitable lands. These opportunities are worth pursuing, especially as a near-term strategy, but they are not large enough to make up the needed emissions reductions if the United States falls short in reducing CO_2 emissions from energy sources.

Acting to reduce GHG emissions in any of these areas will entail costs as well as benefits, but it is difficult to estimate overall economic impacts over time frames spanning decades. While different model projections suggest a range of possible impacts, all recent studies indicate that gross domestic product (GDP) continues to increase substantially over time. Studies also clearly indicate that the ultimate cost of GHG emission reduction efforts depends upon successful technology innovation (Figure S.3).

We thus conclude that there is an urgent need for U.S. action to reduce GHG emissions. In response to this need for action, we recommend the following core strategies to U.S. policy makers:

- Adopt a mechanism for setting an economy-wide carbon-pricing system.
- Complement the carbon price with a portfolio of policies to
  - realize the practical potential for near-term emissions reductions through energy efficiency and low-emission energy sources in the electric and transportation sectors;
  - establish the technical and economic feasibility of carbon capture and storage and new-generation nuclear technologies; and
  - accelerate the retirement, retrofitting, or replacement of GHG emission-intensive infrastructure.
- Create new technology choices by investing heavily in research and crafting policies to stimulate innovation.
- Consider potential equity implications when designing and implementing climate change limiting policies, with special attention to disadvantaged populations.
- Establish the United States as a leader to stimulate other countries to adopt GHG reduction targets.
- Enable flexibility and experimentation with policies to reduce GHG emissions at regional, state, and local levels.
- Design policies that balance durability and consistency with flexibility and capacity for modification as we learn from experience.

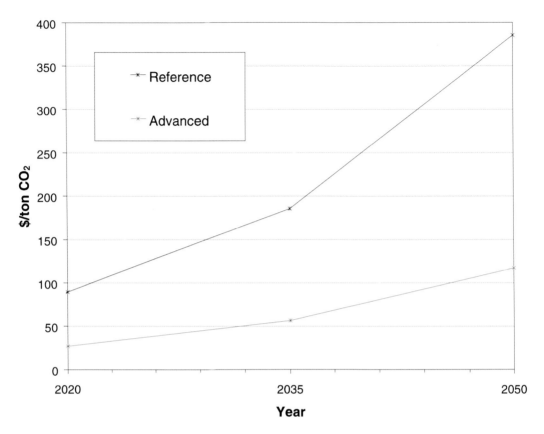

FIGURE S.3 A model projection of the future price of CO_2 emissions under two scenarios: a "reference" case that assumes continuation of historical rates of technological improvements and an "advanced" case with more rapid technological change. The absolute costs are highly uncertain, but studies clearly indicate that costs are reduced dramatically when advanced technologies are available. SOURCE: Adapted from Kyle et al. (2009).

These recommendations are each discussed in greater detail below.

Adopt a mechanism for setting an economy-wide carbon pricing system (see Chapter 4).

A carbon pricing strategy is a critical foundation of the policy portfolio for limiting future climate change. It creates incentives for cost-effective reduction of GHGs and provides the basis for innovation and a sustainable market for renewable resources. An economy-wide carbon-pricing policy would provide the most cost-effective reduction opportunities, would lower the likelihood of significant emissions leakage,[3] and could be designed with a capacity to adapt in response to new knowledge.

Incentives for emissions reduction can be generated either by a taxation system or a cap-and-trade system. Taxation sets prices on emissions and lets quantities of emissions vary; cap-and-trade systems set quantity limits on emissions and let prices vary. There are also options for hybrid approaches that integrate elements of both taxation and cap-and-trade systems.

Any such systems face common design challenges. On the question of how to allocate the financial burden, research strongly suggests that economic efficiency is best served by avoiding free allowances (in cap-and-trade) or tax exemptions. On the question of how to use the revenues created by tax receipts or allowance sales, revenue recycling could play a number of important roles—for instance, by supporting complementary efforts such as research and development (R&D) and energy efficiency programs, by funding domestic or international climate change adaptation efforts, or by reducing the financial burden of a carbon-pricing system on low-income groups.

In concept, both tax and cap-and-trade mechanisms offer unique advantages and could provide effective incentives for emission reductions. In the United States and other countries, however, cap-and-trade has received the greatest attention, and we see no strong reason to argue that this approach should be abandoned in favor of a taxation system. In addition, the cap-and-trade system has features that are particularly compatible with other of our recommendations. For instance, it is easily compatible with the concept of an emissions budget and more transparent with regard to monitoring progress toward budget goals. It is also likely to be more durable over time, since those receiving emissions allowances have a valued asset that they will likely seek to retain.

High-quality domestic and international GHG offsets can play a useful role in lowering the overall costs of achieving a specific emissions reduction, both by expanding the scope of a pricing program to include uncovered emission sources and by offering a financing mechanism for emissions reduction in developing countries (although only for cases where adequate certification, monitoring, and verification are possible, including a demonstration that the reductions are real and additional). Note, however, that using international offsets as a way to meet the domestic GHG emissions budget could ultimately create a more onerous emissions-reduction burden for the countries

---

[3] Emissions leakage refers to the phenomenon whereby controlling emissions within one region or sector causes activity and resulting emissions to shift to another, uncontrolled region or sector.

selling the offsets; absent some form of compensation to ease this burden, seller countries may resist the use of offsets that are counted against their own emission budget.

**Complement the carbon-pricing system with policies to help realize the practi-
cal potential of near-term technologies; to accelerate the retrofit, replacement,
or retirement of emission-intensive infrastructure; and to create new technology
choices (see Chapters 3, 4, and 5).**

Pricing GHGs is a crucial but insufficient component of the effort to limit future climate
change. Because many market barriers exist and a national carbon pricing system
will take time to develop and mature, a strategic, cost-effective portfolio of complementary policies is necessary to encourage early actions that increase the likelihood
of meeting the 2050 emissions budget. This policy portfolio should have three major
objectives, listed below.

1. Realize the practical potential for near-term emission reductions.

End-use energy demand and the technologies used for electricity generation and
transportation drive the majority of U.S. CO_2 emissions. Key near-term opportunities
for emission reductions in these areas include the following:

- *Increase energy efficiency.* Enhancing efficiency in the production and use of
  electricity and fuels offers some of the largest near-term opportunities for
  GHG reductions. These opportunities can be realized at a relatively low marginal cost, thus leading to an overall lowering of the cost of meeting the 2050
  emissions budget. Furthermore, achieving greater energy efficiency in the
  near term can help defer new power plant construction while low-GHG technologies are being developed.
- *Increase the use of low-GHG-emitting electricity generation options, including the
  following:*
  - *Accelerate the use of renewable energy sources.* Renewable energy sources
    offer both near-term opportunities for GHG emissions reduction and po-
    tential long-term opportunities to meet global energy demand. Some re-
    newable technologies are at and others are approaching economic parity
    with conventional power sources (even without a carbon-pricing system
    in place), but continued policy impetus is needed to encourage their de-
    velopment and adoption. This includes, for instance, advancing the devel-
    opment of needed transmission infrastructure, offering long-term stability
    in financial incentives, and encouraging the mobilization of private capital
    support for research, development, and deployment.

- *Address and resolve key barriers to the full-scale testing and commercial-scale demonstration of new-generation nuclear power.* Improvements in nuclear technology are commercially available, but power plants using this technology have not yet been built in the United States. Although such plants have a large potential to reduce GHG emissions, the risks of nuclear power such as waste disposal and security and proliferation issues remain significant concerns and must be successfully resolved.
- *Develop and demonstrate power plants equipped with carbon capture and storage technology.*[4] Carbon capture and storage could be a critically important option for our future energy system. It needs to be commercially demonstrated in a variety of full-scale power plant applications to better understand the costs involved and the technological, social, and regulatory barriers that may arise and require resolution.

• *Advance low-GHG-emitting transportation options.* Near-term opportunities exist to reduce GHGs from the transportation sector through increasing vehicle efficiency, supporting shifts to energy-efficient modes of passenger and freight transport, and advancing low-GHG fuels (such as cellulosic ethanol).

2. Accelerate the retirement, retrofitting, or replacement of emissions-intensive infrastructure.

Transitioning to a low-carbon energy system requires clear and credible policies that enable not only the deployment of new technologies but also the retrofitting, retiring, or replacement of existing emissions-intensive infrastructure. However, the turnover of the existing capital stock of the energy system may be very slow. Without immediate action to encourage retirements, retrofitting, or replacement, the existing emissions-intensive capital stock will rapidly consume the U.S. emissions budget.

3. Create new technology choices.

The United States currently has a wide range of policies available to facilitate technological innovation, but many of these policies need to be strengthened, and in some cases additional measures enacted, to accelerate the needed technology advances. The magnitude of U.S. government spending for (nondefense) energy-related R&D has declined substantially since its peak nearly three decades ago. The United States also

---

[4] Emissions leakage refers to the phenomenon whereby controlling emissions within one region or sector causes activity and resulting emissions to shift to another, uncontrolled region or sector.

lags behind many other leading industrialized countries in the rate of government spending for energy-related R&D as a share of national GDP. While recommendations for desired levels and priorities for federal energy R&D spending are outside the scope of this study, we do find that the level and stability of current spending do not appear to be consistent with the magnitude of R&D resources needed to address the challenges of limiting climate change. In the private sector as well, compared to other U.S. industries, the U.S. energy sector spends very little on R&D relative to income or sales profits.

Research is a necessary first step in developing many new technologies, but bringing innovations to market requires more than basic research. Policies are also needed to establish and expand markets for low-GHG technologies, to more rapidly bring new technologies to commercial scale (especially support for large-scale demonstrations), to foster workforce development and training, and more generally to improve our understanding of how social and behavioral dynamics interact with technology and how technological changes interact with the broader societal goals of sustainable development.

Consider potential equity implications when designing and implementing policies to limit climate change, with special attention to disadvantaged populations (see Chapter 6).

Low-income groups consume less energy per capita and therefore contribute less to energy-related GHG emissions. Yet, low-income and some disadvantaged minority groups are likely to suffer disproportionately from adverse impacts of climate change and may also be adversely affected by policies to limit climate change. For instance, energy-related goods make up a larger share of expenditures in poor households, so raising the price of energy for consumers may impose the greatest burden on these households. Likewise, limited discretionary income may preclude these households from participating in many energy-efficiency incentives. Because these impacts are likely but not well understood, it will be important to monitor the impacts of climate change limiting policies on poor or disadvantaged communities and to adapt policies in response to unforeseen adverse impacts. Some key strategies to consider include the following:

- structuring policies to offset adverse impacts to low-income and other disadvantaged households (for instance, structuring carbon pricing policies to provide relief from higher energy prices to low-income households);
- designing incentive-based climate change limiting policies to be accessible to poor households (such as graduated subsidies for home heating or insulation improvements);

- ensuring that efforts to reduce energy consumption in the transport sector avoid disadvantaging those with already limited mobility; and
- actively and consistently engaging representatives of poor and minority communities in policy planning efforts.

Major changes to our nation's energy system will inevitably result in shifting employment opportunities, with job gains in some sectors and regions but losses in others (i.e., energy-intensive industries and regions most dependent on fossil fuel production). Policy makers could help smooth this transition for the populations that are most vulnerable to job losses through additional, targeted support for educational, training, and retraining programs.

Establish the United States as a leader to stimulate other countries to adopt GHG emissions reduction targets (see Chapter 7).

Even substantial U.S. emissions reductions will not, by themselves, substantially alter the rate of climate change. Although the United States is responsible for the largest share of historic contributions to global GHG concentrations, all major emitters must ultimately reduce emissions substantially. However, the *indirect* effects of U.S. action or inaction are likely to be very large. That is, what this nation does about its own GHG emissions will have a major impact on how other countries respond to the climate change challenge, and without domestic climate change limiting policies that are credible to the rest of the world, no U.S. strategy to achieve global cooperation is likely to succeed. Continuing efforts to inform the U.S. public of the dangers of climate change and to devise cost-effective response options will therefore be essential for global cooperation as well as for effective, sustained national action.

The U.S. international climate change strategy will need to operate at multiple levels. Continuing attempts to negotiate a comprehensive climate agreement under the United Nations Climate Change Convention are essential to establish good faith and to maximize the legitimacy of policy. At the same time, intensive negotiations must continue with the European Union, Japan, and other Organisation for Economic Co-operation and Development (i.e., high-income) countries and with low- and middle-income countries that are major emitters of, or sinks for, GHGs (especially China, India, Brazil, and countries of the former Soviet Union). These multiple tracks need to be pursued in ways that reinforce rather than undermine one another. It may be worthwhile to negotiate sectoral as well as country-wide agreements, and GHGs other than CO_2 should be subjects for international consideration. In such negotiations, the United States should press for institutional arrangements that provide credible assessment and verification of national policies around the world and that help the low- and middle-income countries attain their broader goals of sustainable development.

Competition among countries to take the lead in advancing green technology will play an important role in stimulating emissions-reduction efforts, but strong cooperative efforts will be needed as well. Sustaining large, direct governmental financial transfers to low-income countries may pose substantial challenges of political feasibility; however, large financial transfers via the private sector could be facilitated via a carbon pricing system that allows purchases of allowances or offsets. There is a clear need for support of innovative scientific and technical efforts to help low- and middle-income countries limit their emissions. To provide leadership in these efforts, the United States needs to develop and share technologies that not only reduce GHG emissions but also help advance economic development and reduce local environmental stresses.

Enable flexibility and experimentation with emissions-reduction policies at regional, state, and local levels (see Chapter 7).

State and local action on climate change has already been significant and wide-ranging. For instance, states are operating cap-and-trade programs, imposing performance standards on utilities and auto manufacturers, running renewable portfolio standard programs, and supporting and mandating energy efficiency. Cities across the United States are developing and implementing climate change action plans. Many federal policies to limit climate change will need ongoing cooperation of states and localities in order to be successfully implemented—including, for example, energy-efficiency programs, which are run by localities or state-level programs, and energy-efficiency building standards, which are sometimes enacted statewide and implemented by local authorities. Moreover, states have regulatory capacity (for example, in regulating energy supply and implementing building standards) that may be needed for implementing new federal initiatives.

Subnational programs will thus need to continue playing a major role in meeting U.S. climate change goals. In addition, since climate change limiting policies will continue to evolve, policy experimentation at subnational levels will provide useful experience for national policy makers to draw upon. On the other hand, subnational regulation can pose costs, such as the fact that businesses operating in multiple jurisdictions may face multiple state regulatory programs and therefore increased compliance burdens. Overlap with state cap-and-trade programs can make a federal program less effective by limiting the freedom of the market to distribute reductions efficiently. These costs, however, may be worth the benefits gained from state and local regulatory innovation; regardless, they illustrate the trade-offs Congress will have to consider when deciding whether to preempt state action.

Thus, a balance must be struck to preserve the strengths and dynamism of state and local actions, while tough choices are made about the extent to which national policies should prevail. In some instances, it may be appropriate to limit state and local authority and instead mandate compliance with minimum national standards. But Congress could promote regulatory flexibility and innovation across jurisdictional boundaries when this is consistent with effective and efficient national policy. To this end, we suggest that Congress avoid punishing or disadvantaging states (or entities within the states) that have taken early action to limit GHG emissions, avoid preempting state and local authority to regulate GHG emissions more stringently than federal law without a strong policy justification, and ensure that subnational jurisdictions have sufficient resources to implement and enforce programs mandated by Congress.

Design policies that balance durability and consistency with flexibility and capacity for modification as we learn from experience (see Chapter 8).

The strategies and policies outlined above are complex efforts with extensive implications for other domestic issues and for international relations. It is therefore crucial that policies be properly implemented and enforced and be designed in ways that are durable and resistant to distortion or undercutting by subsequent pressures. At the same time, policies must be sufficiently flexible to allow for modification as we gain experience and understanding (as discussed earlier in this Summary). Transparent, predictable mechanisms for policy evolution will be needed.

There are inherent tensions between these goals of durability and adaptability, and it will be an ongoing challenge to find a balance between them. Informing such efforts requires processes for ensuring that policy makers regularly receive timely information about scientific, economic, technological, and other relevant developments. One possible mechanism for this process is a periodic (e.g., biennial) collection and analysis of key information related to our nation's climate change response efforts. This effort could take the form of a "Climate Report of the President" that would provide a focal point for analysis, discussion, and public attention and, ideally, would include requirements for responsible implementing agencies to act upon pertinent new information gained through this reporting mechanism.

CHAPTER ONE

Introduction

Climate change, driven by the increasing concentration of greenhouse gases (GHGs) in the atmosphere, poses serious, wide-ranging threats to human societies and natural ecosystems around the world. While many uncertainties remain regarding the exact nature and severity of future impacts, the need for action seems clear. In the legislation that initiated our assessment of America's climate choices, Congress directed the National Research Council to "investigate and study the serious and sweeping issues relating to global climate change and make recommendations regarding the steps that must be taken and what strategies must be adopted in response to global climate change." As part of the response to this request, the America's Climate Choices Panel on Limiting the Magnitude of Future Climate Change was charged to "describe, analyze, and assess strategies for reducing the net future human influence on climate, including both technology and policy options, focusing on actions to reduce domestic greenhouse gas (GHG) emissions and other human drivers of climate change, but also considering the international dimensions of climate stabilization" (see full statement of task in Appendix B). In other words, this report examines the questions, "What are the most effective options to help reduce GHG emissions or enhance GHG sinks?" and "What are the policies that will help drive the development and deployment of these options?"

CONTEXT AND PURPOSE OF THIS REPORT

Devising strategies to limit future climate change involves an extensive and complex set of issues, and national leaders will be required to make difficult choices in responding to these issues. Increasing understanding of the risks and challenges involved in limiting the magnitude of climate change compels this panel to urge early, aggressive, and concerted actions to reduce emissions of GHGs. Although many technology and policy responses are available, the practical challenges to realizing their potential are immense. A large-scale national commitment that both generates and is underpinned by international cooperation is crucial if the risks of global climate change are to be substantially curtailed. Because action to initiate these efforts is urgently needed, our principal focus has been on steps that can and should be taken now. Specifically, we focus particular attention on strategies to

- *Reduce concentrations of GHGs in the atmosphere as the principal means to limit the magnitude of climate change.* This is primarily a challenge of reducing net GHG emissions (directly, or possibly through enhanced sequestration), but it could also encompass strategies to remove GHGs directly from the atmosphere.
- *Promote options and policies that appear to be technically and economically feasible now or could become feasible in the near term.* However, the report also identifies other strategies that may play an important role in the future but whose potential cannot be reliably estimated, and we encourage policies that can be adopted now to accelerate these future innovations.

Although the urgent need for action is real, many relevant efforts are already under way. For example, 24 U.S. states and more than 1,000 U.S. cities have adopted some form of targets for limiting GHG emissions. The judicial system is also playing an increasingly important role, and two recent lawsuits may result in GHG emissions reduction through administrative or judicial action. As a result of *Massachusetts v. EPA*, the Environmental Protection Agency has issued a finding under the Clean Air Act that GHGs are endangering public health and welfare, which may form the basis for extensive regulation of GHGs under the Act. In *Connecticut v. American Electric Power*, the Second Circuit allowed a public nuisance case by several states and nonprofit organizations to go forward against the country's largest utilities, alleging that past and ongoing CO_2 emissions are contributing to global warming. If the plaintiffs ultimately succeed, the defendants presumably could be required to reduce their emissions. Furthermore, during the course of this study, Congress and the Obama Administration have been considering legislation to establish a national program for limiting future climate change.

It is not our intent to comment on these actions or to evaluate the policies they embrace. Rather, we regard them as the necessary beginning of an ongoing process of developing a coherent national policy for limiting future climate change. Although we are optimistic that the United States can meet the challenges associated with limiting future climate change, we are also convinced that this can only be done as a long-term evolving process. Accordingly, our study develops two major themes in addition to the need for urgent action:

- *Provide a national framework of strategies and policies to limit climate change, now and in the future.* National policy goals must be implemented through the actions of the private sector, other levels of government, and individuals. The role of the federal government is to provide strong leadership and help shape the landscape in which all of these actors make decisions. Although this report focuses on policy at the national level, our aim is to suggest a policy framework within which all actors can work effectively toward a shared national goal. In addition, we consider how U.S. actions can provide incentives for effective action on climate change by other countries.

- *Chart a course for managing the policy process over the coming decades.* It is inevitable that policies put in place today will need to be adjusted over time as new scientific information changes our understanding of the magnitude and nature of climate change. Moreover, the experience of implementing even well-conceived policies will no doubt produce unexpected difficulties (and, one hopes, surprising successes) to which future policy should adapt. And finally, technological innovation—or lack thereof—will alter the strategies available for limiting future climate change. It will be essential for policy makers to respond regularly to new knowledge about science, technology, and policy if we are to address climate change successfully.

Because of the considerable complexities involved in climate change limiting policy, existing research and analysis do not always point to unequivocal recommendations. Where research clearly shows that certain policy design options are particularly effective, we recommend specific goals for the evolving policy portfolio. In other cases, however, we simply examine the range of policy choices available to decision makers and identify sources of further information on these choices. Relatively little research seems to be available on a few fundamentally important issues. For example, the problem of designing a durable yet adaptable policy framework to guide actions over decades is not well understood. In these cases, we raise the issue and point to the need for further analysis or research.

PRINCIPLES TO GUIDE CLIMATE CHANGE LIMITING POLICY AND STRATEGY

To provide structure and rigor to the process of evaluating alternative climate change limiting policies and strategies, the panel developed a set of guiding principles, selected after reviewing and debating nearly a dozen examples of principles for climate policy that have been proposed by nongovernmental organizations, congressional leaders, and others. The principles are intended to be enduring—reflecting nonpartisan, cross-generational, and pluralistic values. We acknowledge that an inherent tension exists among some of these principles, and thus it may not always be possible to satisfy them all simultaneously. The principles are not themselves policy prescriptive, nor do they represent a set of specific goals or desired outcomes. Rather, they were used as a framing exercise to help us identify priority recommendations. The principles are as follows:

- *Environmental Effectiveness*: Set short- and long-term emissions-reduction targets that are consistent with meeting environmental goals.
- *Cost Effectiveness*: Achieve emission reductions at lowest possible costs, paying special attention to costs of delay.
- *Innovation*: Stimulate entrepreneurial capacity to advance technologies and strategies for reducing GHG emissions.
- *Equity and Fairness*: Strive for solutions that are fair among people, regions, nations, and generations, taking into account existing global disparities in consumption patterns and capacity to adapt. Do not penalize those who have taken early action to reduce emissions.
- *Consistency*: Ensure that new legislation is consistent with existing legislation to avoid delay and confusion.
- *Durability*: Sustain action over several decades, sending clear signals to investors, consumers, and decision makers.
- *Transparency*: Ensure that goals and policies, and their rationales, are clear to the public; establish clear benchmarks to assess and publicly report on progress.
- *Adaptability*: Review and adjust polices in response to evolving scientific information and socioeconomic and technological changes.
- *Global Participation*: Enhance engagement with other countries to cooperate in achieving climate change limiting goals. The climate challenge requires a global solution.
- *Regional, State, and Local Participation*: Encourage and support regional, state, local, and household action in ways that are consistent with national goals and policies.

Many of these principles are discussed further in the subsequent chapters.

ORGANIZATION OF THE REPORT

The balance of this report is organized as follows:

- **Chapter 2** characterizes the challenge of limiting climate change by examining historical, current, and likely future GHG emission sources, and it examines representative targets and pathways for reducing U.S. domestic emissions.
- **Chapter 3** examines the key strategies for limiting atmospheric concentrations of GHGs. To the extent possible, we assess how these strategies might attain the targets and pathways developed in Chapter 2.

- **Chapter 4** analyzes federal policies for implementing the key emission reduction strategies discussed above, and, where an adequate research basis exists, recommends critical policy goals.
- **Chapter 5** discusses policies for promoting the technological innovation needed to develop strategies that are less costly and more effective than those in hand today.
- **Chapter 6** examines how climate change limitation strategies interact with other important national policy goals, including social equity concerns, economic development threats and opportunities, energy security, and protection of air and water quality.
- **Chapter 7** discusses policies for assuring a multilevel response to the climate problem. In particular, we consider how U.S. national policy affects international incentives and institutions, as well as state and local actions.
- **Chapter 8** explores the concepts of policy durability and evolution as the basis for guiding the policy process over the coming decades.

CHAPTER TWO

Goals for Limiting Future Climate Change

The purpose of this report is to suggest strategies for limiting the magnitude of future climate change. Although limiting climate change is a global issue, our assignment is to recommend domestic strategies and actions. A goal for reducing U.S. greenhouse gas (GHG) emissions is therefore needed as a basis for designing domestic policies, for evaluating their feasibility, and for monitoring their effectiveness. In this chapter, we examine how such goals can be set. We first assess future domestic and global "reference" GHG emission scenarios, that is, scenarios that would result without new policies to limit future emissions. We then outline the key steps involved in formulating goals for policies that limit climate change and recommend that goals be framed in terms of limits on cumulative domestic GHG emissions over a specified time period. In choosing a specific goal for the United States, policy makers will have to deal not only with scientific uncertainties but also with ethical judgments. Because the judgments can be informed but not fully answered by science, we do not attempt to recommend a specific U.S. emissions budget. However, in order to have a basis for identifying and evaluating policy recommendations, we have used recent modeling studies (most notably the EMF22 study[1]) to suggest a plausible range for a domestic GHG emissions budget. We then examine options for global and U.S. emissions-reduction goals, respectively, and finally we examine some potential economic impacts of these emission reduction goals.

REFERENCE U.S. AND GLOBAL EMISSIONS

The most recent data for U.S. GHG emissions (for the year 2007) show a total of 7,150 million CO_2-equivalent tons (Mt CO_2-eq).[2] Over 85 percent of this total is CO_2 emis-

---

[1] Stanford University Energy Modeling Forum (EMF22; see *http://emf.stanford.edu/research/emf22/* and Clarke et al., 2009). Although other models can be used, as explained in Box 2.2, we believe that EMF22 is particularly useful for our purposes and that the insights from EMF22 are consistent with the broader literature.

[2] A common practice is to compare and aggregate emissions among different GHGs by using global warming potentials (GWPs). Emissions are converted to a CO_2 equivalent (CO_2-eq) basis using GWPs as published by the Intergovernmental Panel on Climate Change (IPCC). GWPs used here and elsewhere are calculated over a 100-year period, and they vary due to the gases' ability to trap heat and their atmospheric

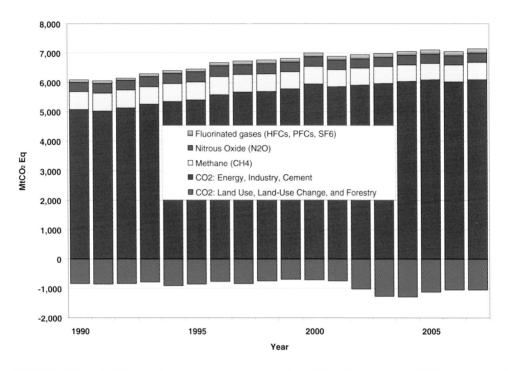

FIGURE 2.1 Historic U.S. greenhouse gas emissions and sinks. CO_2 is the dominant GHG, but the contributions from other GHGs are not insignificant. SOURCE: EPA (2009).

sions, and roughly 94 percent of the CO_2 emissions comes from combustion of fossil fuel (with most of the rest arising from industrial processes such as cement manufacturing). Methane (CH_4) makes up about 8 percent of total emissions, nitrous oxide (N_2O) about 4 percent, and the fluorinated gases (hydrofluorocarbons [HFCs], perfluorocarbons [PFCs], SF_6) about 2 percent. There is also a net CO_2 sink (removal from the atmosphere) from land-use and forestry activities, estimated at 1,063 Mt CO_2 in 2007. Between 1990 and 2007, total U.S. GHG emissions have risen by 17 percent, with a relatively steady annual average growth of 1 percent per year. Figure 2.1 illustrates these trends (EPA, 2009).

The main drivers of GHG emissions include population growth and economic activity, coupled with the intensity of energy use per capita and per unit of economic output. Figure 2.2 shows that U.S. primary energy use has continued to grow over the

lifetime, compared to an equivalent mass of CO_2. Although GWPs were updated in IPCC (2007b), emission estimates in this report continue to use GWPs from IPCC (1995), to be consistent with international reporting standards under the United Nations Framework Convention on Climate Change (UNFCCC).

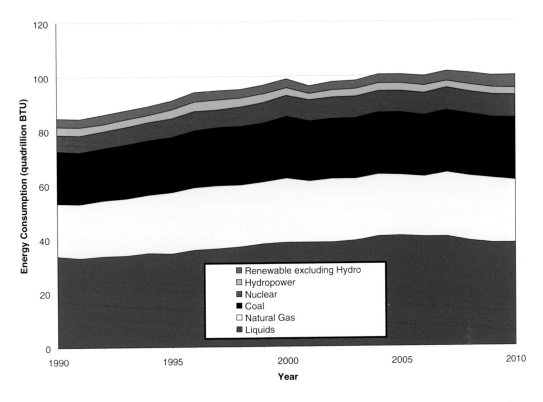

FIGURE 2.2 U.S. primary energy use, 1990 to 2010. Fossil fuels are the dominant energy source over this period. "Liquids" refers petroleum products including gasoline, natural gas plant liquids, and crude oil burned as fuel, but it does not include the fuel ethanol portion of motor gasoline. SOURCE: EIA (2009).

period of 1990 to 2010, although at a decreasing rate: Total energy consumption has grown at a slower pace than economic output and population. This slower growth in energy consumption stems from structural changes in the U.S. economy (e.g., the shift to a more service-oriented economy) as well as increasing energy efficiency per unit of economic output. Trends in GHG emissions are closely associated with energy consumption. Figure 2.3 compares growth in GHG emissions with growth in primary energy use, population, and economic output in the United States. Since 1990, the U.S. economy has doubled in size while the population has grown about 20 percent and energy use and GHG emissions have grown 10 to 15 percent (EPA, 2009). Recent government projections out to 2030 are for economic growth to continue along historic rates, outpacing growth in energy use and GHG emissions because of the reduced energy intensity of the economy.

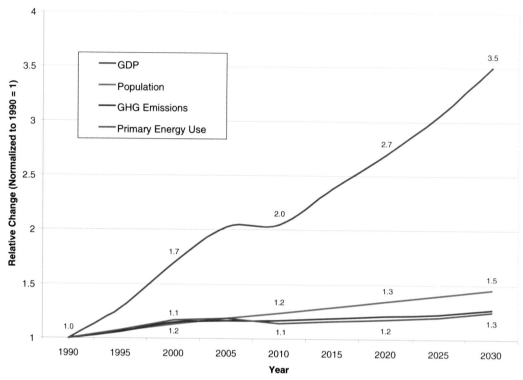

FIGURE 2.3 Historical trends and projected future trends in U.S. GHG emissions (including CO_2, CH_4, N_2O, HFCs, PFCs, and SF_6, but excluding net land-use emissions) and indices of key emission drivers: population, primary energy use, and economic growth (gross domestic product [GDP]). GHG emissions have risen roughly in concert with growth in energy use and population, but substantially slower than the rate of overall economic growth. The base year for calculating the indices is 1990. GDP estimates used to calculate the GDP index are based on real 2005 U.S. dollars. SOURCES: Historic data are from EPA (2009) and CEA (2009); projected data are from the ADAGE model (EPA, 2009).

Figure 2.4 provides a range of recent GHG emission scenarios from various models (EIA, 2009; Fawcett et al., 2009) assuming no mitigation policies are in place. The chart shows that emissions in 2030 range from 7,100 to 8,400 Mt CO_2-eq and in 2050 range from 8,100 to 10,900 Mt CO_2-eq. Variations in projections are the result of varying assumptions of economic growth, energy efficiency, and the deployment of energy technologies (all in the absence of national GHG emissions-reduction policies). For example, MIT's EPPA model assumes an annual GDP growth rate of 2.5 percent per year from 2005 to 2050 (Paltsev et al., 2009), while the Electric Power Research Institute's (EPRI's) MERGE model assumes lower annual growth rates starting at 2.2 percent through 2020 and declining to 1.3 percent through 2050 (Blanford et al., 2009). In addition, the MERGE model assumes a movement away from oil and toward more electric generation as a

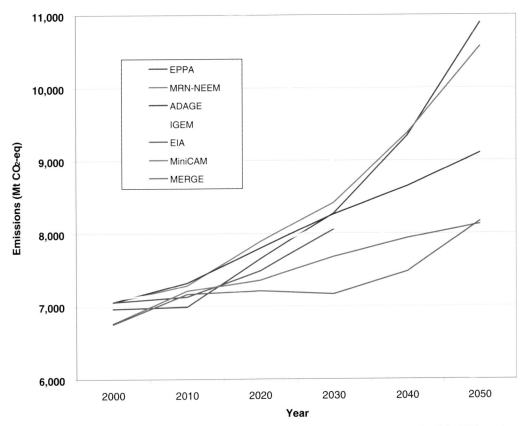

FIGURE 2.4 Reference ("no policy") GHG emission scenarios including CO_2, CH_4, N_2O, HFCs, PFCs, and SF_6. These scenarios include source emissions across many sectors but exclude net emissions from land use–related carbon sequestration. Despite the wide range of outcomes among the different projections, they all show increasing emissions over time. The Applied Dynamic Analysis of the Global Economy (AD-AGE) model results are the same as the GHG emissions line used in Figure 2.3. SOURCES: Adapted from Fawcett et al. (2009) and EIA (2009).

share of energy use. The combined assumptions of lower economic growth and less carbon-intensive energy use produce lower GHG emissions in the MERGE reference projection. Recognizing the inherent uncertainty associated with making long-term projections, two key insights emerge from the reference projections:

- In the absence of emission mitigation policies, annual U.S. GHG emissions will continue to increase out to 2050 (even as the energy intensity of the economy declines).
- The earlier that measures are taken to influence the trajectory of emissions, the more long-term emissions can be reduced.

Similar to U.S. projections, global projections of GHG emissions are determined by the dynamic interaction of key emissions drivers, most notably population and economic growth, as well as the intensity of energy use (per capita and per unit of economic output) and technological change. Projections of GHG drivers, emissions, and concentrations are taken from the recent EMF22 study (Clarke et al., 2009). Figure 2.5 provides recent population projections from models participating in the EMF22 study. The mean of the global population estimates for 2010 is about 7 billion people. Global population projections for 2050 have a mean of about 9 billion people. In all of the employed reference models, population growth rates are projected to slow toward the end of the 21st century, producing a mean projection of 9.5 billion people in 2100 but a wide variability among the models (from 8.7 to 10.5 billion people). Such variability among models can be expected, since long-range global population projections embody different assumptions about regional population trends. For example, trends

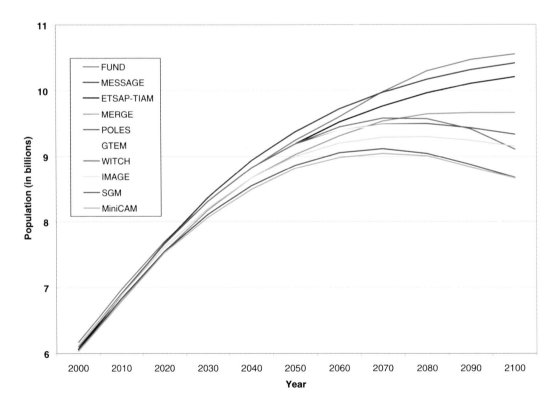

FIGURE 2.5 Reference global population projections from the models used in the EMF22 study. The divergence among the different model projections grows over time. SOURCE: L. Clarke, Pacific Northwest National Laboratory (PNNL).

in sub-Saharan Africa, the Middle East and North Africa, and the East Asia regions are driven by changes in lower-than-expected fertility rates based on recent data. In the Organisation for Economic Co-operation and Development (OECD) region, by contrast, recent projections are somewhat higher than previous estimates mainly due to changes in assumptions regarding migration and more optimistic projections of future life expectancy.

The reference projections for global primary energy consumption from the EMF22 study are shown in Figure 2.6. The range of global energy production in 2050 is between 790 and 1,115 exajoules[3] with a mean value of about 890 exajoules. The rates of growth in primary energy consumption are greater than population growth, leading to an even greater per capita energy use out to 2100. By the end of the century, global energy production is projected to be between 2.5 and 3.5 times greater than today's levels. The key reasons for differences in total primary energy projections include assumptions about population and economic growth; improvement in energy intensity, that is, the relationship between energy consumption and economic output over time; the abundance of different fuels and their relative prices; and the availability and deployment of energy technologies. For example, a scenario that projects more coal use will result in more CO_2 emissions than one where natural gas and renewable energy represent a larger share of total energy consumption.

Figure 2.7 shows global projections of fossil and industrial CO_2 emissions and the CO_2-eq concentrations[4] from all Kyoto Protocol gases (CO_2, CH_4, N_2O, HFCs, PFCs, and SF_6) from the EMF22 reference (no policy) scenarios (Clarke et al., 2009). Reference projections of global GHG emissions and concentrations highlight the fact that global emissions and concentrations will increase substantially over the century, with attendant changes in the global climate. There is a wide spread in emissions projections, resulting from many of the same uncertainties that are reflected in the primary energy projections, along with uncertainties about the development and deployment of low-carbon energy technologies without mitigation policy. Regardless, the projections all indicate upward trends. By the end of the century, the range of CO_2-eq concentrations spans from two times to almost four times today's levels. (That is, concentrations in

---

[3] Primary energy is energy contained in raw fuel that has not been subjected to any conversion or transformation process. One exajoule = 10^{18} joules. A joule is the work required to continuously produce one watt of power for one second.

[4] CO_2-eq (CO_2 equivalent) concentration is defined as a multi-GHG concentration that would lead to the same impact on the Earth's radiative balance as a concentration of CO_2 only (IPCC, 2007b). See Box 2.1 for further discussion.

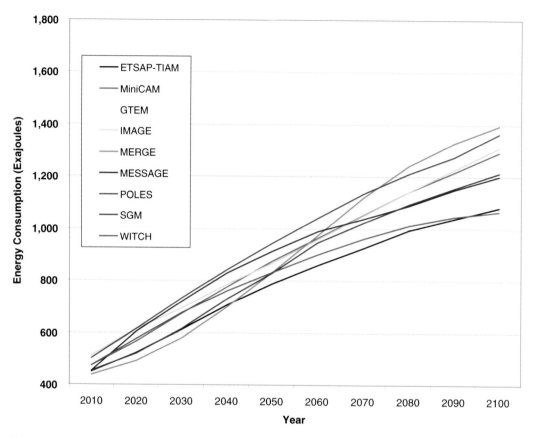

FIGURE 2.6 Reference global primary energy consumption projections from the models used in the EMF22 study. Note that all estimates project considerable growth in energy production over the course of the century. SOURCE: Adapted from Clarke et al. (2009).

2050 range between 800 and 1,500 ppm[5] CO_2-eq, in contrast to today's concentrations of roughly 440 ppm CO_2-eq.)

Although the high-income (OECD) countries are currently the largest contributors to cumulative GHG emissions, emissions from rapidly growing low- and middle-income countries (e.g., Brazil, China, and India) are projected to grow more quickly than those of high-income countries. Figure 2.8 shows historical and projected contributions to global emissions out to 2100 from several sources. In all the projections, the balance of cumulative GHG contributions shifts from the high-income to the low- and middle-income countries through 2050; in second half of the 21st century, the low- and middle-

_____

[5] Parts per million (by volume, sometimes abbreviated as ppmv).

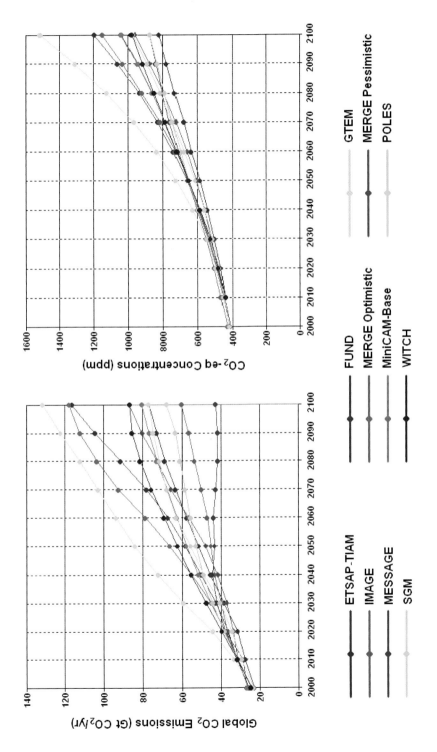

FIGURE 2.7 Projections of global CO_2 emissions (from fossil and industrial sources) and CO_2-equivalent concentrations (CO_2-eq) in the absence of efforts to address climate change, from the EMF22 study. The left panel shows global (fossil and industrial) CO_2 emissions from across models. The right panel shows the CO_2-eq concentrations, including all the Kyoto Protocol gases, for the same corresponding scenarios. Model projections vary, but all show increasing emissions and concentrations over time. See Box 2.1 for definition of CO_2-eq concentrations. SOURCE: L. Clarke, PNNL.

income countries are projected to account for the bulk of cumulative global GHG emissions.

Together, these factors frame the international context in which the United States will need to decide on its domestic emissions-reduction goals, and also on its support for and involvement in international actions. Even today, as one of the largest individual GHG emitters, the United States cannot substantially reduce global emissions through unilateral action. With its shrinking relative contribution to global emissions, unilateral action by the United States would be decreasingly effective from a quantitative perspective. In some sense, then, a primary role of U.S. action in climate change is to provide global leadership and to motivate effective international action. See Chapter 7 for a fuller discussion of these issues.

SETTING CLIMATE CHANGE LIMITING GOALS

International policy goals for limiting climate change were established in 1992 under the UNFCCC, in which the United States and more than 190 other nations set the goal of "stabilization of GHG concentrations in the atmosphere at a level that would prevent dangerous anthropogenic interference with the climate system." Subsequent scientific research has sought to better understand and quantify the links among GHG emissions, atmospheric GHG concentrations, changes in global climate, and the impacts of those changes on human and environmental systems. Based on this research, many policy makers in the international community recognize limiting the increase in global mean surface temperature to 2°C above preindustrial levels as an important benchmark; this goal was embodied in the Copenhagen Accords, at a 2009 meeting of the G-8, and in other policy forums.

Although these temperature and concentration goals are essential metrics for limiting global climate change over time, they are not sufficient to guide near-term, domestic policy goals. Policy requires a goal linked to outcomes that domestic action can directly affect and that can be measured contemporaneously. Global temperature and concentration goals lack this attribute, because they are the consequence of global, and not just domestic, actions to limit GHG emissions. To avoid this problem, a limit on cumulative emissions from domestic sources, measured in physical quantities of GHGs allowed over a specified time period, is in the panel's view a more useful domestic policy goal. Policy can affect emissions directly, and actual emissions can be measured reasonably accurately on a current basis.

Calculating the U.S. emissions budget is conceptually straightforward, but it involves a number of uncertainties and judgments that are complex and potentially contro-

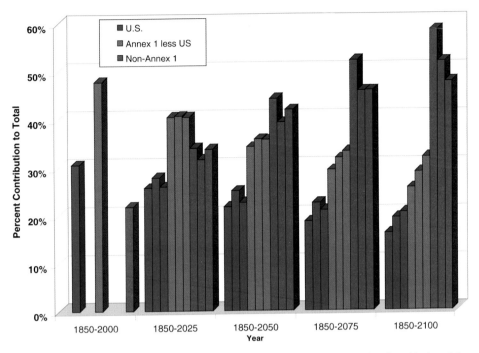

FIGURE 2.8 Historical and future contributions to global CO_2 emissions from fossil and industrial sources (does not include net CO_2 emissions from land use). Annex I and non-Annex I refer to high-income and low- and middle-income groups of countries, respectively, under the UNFCCC. The three bars for each color within each time period represent emissions projections from three models used in the U.S. Climate Change Science Program (CCSP) studies: MIT's EPPA model, EPRI's MERGE model, and PNNL's MiniCAM model. Note that the United States and other high-income countries have had the dominant share of emissions historically, but this share is projected to decrease over time. SOURCES: Historical estimates from Climate Analysis Indicators Tool, Version 6 (WRI, 2009); projections are from U.S. CCSP 2.1a (Clarke, 2007).

versial. To systematically derive an emissions budget from global temperature and concentration goals requires establishing three crucial links:

1. a target global atmospheric GHG concentration that is consistent with an acceptable global mean temperature change,
2. a global emissions budget that is consistent with a target atmospheric GHG concentration, and
3. an allocation to the United States of an appropriate share of the global emissions budget.

According to the most recent assessment of the IPCC, the best estimate of global mean temperature increase over preindustrial levels resulting from stabilizing atmospheric GHG concentrations over the long term at 450 CO_2-eq is 2°C;[6] the best estimate of temperature increase resulting from stabilizing atmospheric GHG concentrations at 550 CO_2-eq is 3°C (Meehl and Stocker, 2007). However, as discussed in Box 2.1, linking global temperature change with a target GHG concentration involves a number of physical processes that are not fully understood; thus, there is substantial uncertainty surrounding this linkage. For example, a 450 ppmv CO_2-eq concentration could be associated with temperature change below 2°C or well above 2°C (Meehl and Stocker, 2007). Future research may change prevailing views about desired temperature change limits or the atmospheric GHG concentration for achieving these temperature limits. Nonetheless, for the purposes of this report, we use these two concentrations (450 and 550 CO_2-eq) as guideposts for considering global emissions budgets and U.S. emissions allocations. The following two sections address these linkages.

The next step of the goal-setting process—allocating an appropriate share of the global budget to the United States—is also a matter of considerable uncertainty and judgment. While science may inform this judgment, it depends mostly on ethical and political considerations (e.g., debates over whether allocation criteria should be based on economic efficiency criteria or on "fairness" criteria). For these reasons, we do not attempt to recommend a specific domestic emissions budget. We do, however, draw upon the work of the EMF22 modeling exercises (see Box 2.2), which identifies specific domestic emissions budget scenarios that are linked to various global action cases. We have adopted the EMF22 results as representative emission targets, both to illustrate an approximate range into which a U.S. budget might fall and to provide some sort of benchmark for evaluating the technical feasibility of reaching a budget target. We discuss the EMF22 results in the following section.

GLOBAL EMISSION TARGETS

EMF22 modeled reference scenarios to provide a range of possible paths of emissions growth under the assumption of no new climate policies. Figure 2.7 shows this range (based on forcing from the Kyoto gases only) and shows that the 450 and 550 CO_2-eq limits will soon be exceeded without aggressive and immediate new emission reduction efforts. To explore what sort of actions might be required in this regard, EMF22 evaluated 10 international climate-action cases, based on combinations of three key elements of a long-term GHG emissions-reduction strategy: (1) long-term concentra-

---

[6] This temperature result is based on using the IPCC "best estimate" climate sensitivity of 3°C and the relationship between radiative forcing and temperature found in the IPCC's Fourth Assessment Report (IPCC, 2007b).

tion goals, using either 450, 550, or 650 CO_2-eq; (2) whether the concentration goal could be exceeded before the end of the 21st century (overshoot or not-to-exceed); and (3) the assumption of international participation—either full participation, meaning all countries undertake emissions-reduction efforts starting in 2012, or delayed participation, meaning low- and middle-income countries do not begin emissions-reduction efforts until 2030 or beyond. Figure 2.9 presents the global fossil and industrial CO_2 emissions in 2050, relative to 2000 levels, resulting from the scenarios analyzed in the EMF22 international study. In each of the cases, resulting emissions are well below what they would be without the implementation of climate change limiting polices.

It is important to keep in mind that these are simply scenarios, and they do not represent all the ways that particular goals could be achieved. In particular, the models are based on a "least-cost" approach to emissions-reduction efforts, which assumes that all nations undertaking emissions-reduction efforts do so in the most cost-effective manner (i.e., emissions reductions undertaken where, when, and how they will be least expensive), under the equivalent of a frictionless cap-and-trade system with full international trading covering all Kyoto gases. One implication of this least-cost approach is that some sectors or countries will undertake more or less mitigation than others. In addition, the models assume transparent markets, no transaction costs, and perfect implementation of emissions-reduction measures throughout the 21st century. At the same time, the delayed participation scenarios assume a highly inefficient architecture in which many of the low- and middle-income countries undertake no emissions reductions before 2030 or even 2050.

The more stringent climate-action cases (including the 450 ppm CO_2-eq cases and the 550 ppm CO_2-eq case without immediate, full global participation) could not be represented in some or all models. These are shown at the bottom of Figure 2.9.[7] This does not necessarily imply that these climate-action cases are impossible in some absolute sense; rather, it is one of several indicators of the challenges of meeting the particular goals under the particular constraints that are represented by a given climate-action case. In general, the difficulty of successfully modeling scenarios increased with the stringency of the long-term goal, with the requirement not to exceed the goal dur-

---

[7] More specifically, "could not be represented" means that (1) the climate-action case was physically infeasible according to the model because the radiative forcing target was exceeded prior to the initiation of mitigation actions in low- and middle-income countries; (2) the initial carbon price exceeded $1,000 per ton CO_2-eq in 2012; or (3) the model could not be solved for the particular climate-action case due to failures in the solution mechanism at higher CO_2-eq prices, constraints that hold back the rate of change in key sectors, or CO_2-eq price limits prescribed in the model.

BOX 2.1
Key Uncertainties in Setting Goals for Limiting Climate Change

The report *ACC: Advancing the Science of Climate Change* (NRC, 2010a) provides a detailed discussion of the scientific uncertainties involved in setting targets for limiting climate change. Here we provide a brief overview of some key uncertainties that affect the results presented later in this chapter.

Climate sensitivity. The quantitative relationship between long-term temperature changes and atmospheric GHG concentrations is very difficult to specify, due primarily to the large uncertainty of climate sensitivity. Climate sensitivity is typically defined as the global mean equilibrium temperature response to a doubling of CO_2 concentrations. IPCC (2007b) indicated that climate sensitivity is *likely* to be in the range of 2°C to 4.5°C with a best estimate of about 3°C. It is *very unlikely* to be less than 1.5°C,[1] and values substantially higher than 4.5°C cannot be excluded. Some recent studies have in fact suggested that much higher climate sensitivity values are possible (Hansen et al., 2008; Sokolov et al., 2009). This uncertainty indicates that relying only on the best estimate of 3°C may not be a prudent risk-management strategy.

Because of these uncertainties, the temperature-concentration relationship is often given in probabilistic terms. As noted earlier, recent research (Meehl and Stocker, 2007; Wigley et al., 2009) indicates that limiting global GHG concentrations to around 450 ppm CO_2-eq over the long term would result in a 2°C temperature change using a climate sensitivity of 3°C. The best-estimate increase in temperature for a long-term concentration of 550 ppm CO_2-eq is 3°C (or 2°C, if the climate sensitivity is at the lowest end of the IPCC range). However, there is significant uncertainty around these point estimates. For example, a study using three models from the EMF22 exercise finds that the probability of staying below 2°C for several "overshoot" scenarios (see below) leading to 450 ppmv CO_2-eq ranges between 24 and 72 percent, depending on the degree of overshoot and on which probability distribution from the literature is used (Krey and Riahi, 2009).

A related consideration is that, due to time lags in the climate system, one might allow actual concentrations to temporarily and modestly exceed (or "overshoot") 450 ppm CO_2-eq while allowing temperature to remain below the 2°C temperature goal. However, an overshoot scenario entails ad-

---

[1] In the IPCC treatment of uncertainty, the following likelihood ranges were used to express the assessed probability of occurrence: likely, >66 percent; very unlikely, <10 percent.

---

ing the century, and with delays in full global participation. For instance, only 2 of the 14 participating models[8] were able to produce scenarios that attained the 450 CO_2-eq goal without immediate, full global participation, and only then if an overshoot trajectory to the goal was allowed. Without the option to overshoot the goal, and with

---

[8] Note that four groups used more than one version of their model in the study. Accounting for these additional analyses, there are 14 models that attempted the 450 CO_2-eq goal without immediate, full global participation.

ditional climate risks that depend on how the climate system responds to concentrations above 450 ppm CO_2-eq. For instance, this could send the climate system over critical thresholds (e.g., irreversible drying of the subtropics, melting large glaciers, and raising sea levels); once such thresholds are crossed, reducing CO_2-eq concentrations may be ineffective for bringing the climate system back to a particular state (Solomon et al., 2009). Finally, allowing emissions to follow an overshoot pathway in the near term leaves open the possibility that, once the concentration target is exceeded, the necessarily steeper emissions declines later in the century may never materialize.

Radiative forcing and CO_2-eq concentrations. Radiative forcing is a measure of impact on the Earth's radiative balance from changes in concentrations of key substances such as GHGs and aerosols. It is generally expressed in terms of watts per meters squared (W/m^2) but can also be expressed in terms of CO_2-equivalent (CO_2-eq) concentrations, that is, the concentrations of CO_2 only that would lead to the same impact on the Earth's radiative balance (Ramaswamy et al., 2001). Radiative forcing agents in the atmosphere that are most relevant to considerations of future climate change include the "Kyoto" gases (those included in the Kyoto Protocol: CO_2, CH_4, N_2O, HFCs, PFCs, and SF_6); CFCs and other ozone-depleting substances covered by the Montreal Protocol; tropospheric ozone; different types of aerosols including sulfates, black carbon, and organic carbon; and land use changes that affect the reflectivity of the Earth's surface.

Many global emissions scenarios consider only CO_2 or only the Kyoto gases. The results of analyses that include only Kyoto gas forcing differ from those that include "full forcing," primarily due to the impact of aerosols. The aerosol influence is complex; some types such as black carbon exert positive forcing (warming), while other types such as sulfates exert negative forcing (cooling). It is estimated that, overall, aerosol-related cooling influences currently lower total forcing by roughly an equivalent of 50 ppm CO_2 (Forster and Ramaswamy, 2007). Many studies indicate that the aerosol influence will attenuate over the coming century, however, particularly if strong climate change limiting policies are enacted. This is because aerosol emissions from fossil fuel combustion are expected to be significantly reduced, both as an indirect result of GHG mitigation efforts and as a direct result of concerns over health impacts. For example, in scenarios from the EMF22 models that most comprehensively considered full forcing, it was found that, by the end of the century, Kyoto-only CO_2-eq concentrations were roughly equal to full-forcing concentrations.

delays in global participation, no models could produce the scenario that met 450 ppm CO_2-eq by 2100.

The EMF22 results indicate that atmospheric GHG concentrations can be kept below 450 ppm CO_2-eq only if the United States and other high-income countries, along with China, India, and many other low- and middle-income countries around the world, take aggressive actions to reduce emissions starting within the next few years. This would represent a dramatic change from recent trends across the globe. If the major

---

BOX 2.2
Our Use of EMF22

To identify plausible goals for a U.S. GHG emissions budget, and to evaluate strategies for meeting this budget, we have relied largely on the work of Energy Modeling Forum Study 22 (EMF22). EMF22 included two components that are relevant here: an international component that engaged 10 of the world's leading integrated assessment models to assess global climate regimes (Clarke et al., 2009) and a U.S. component that engaged six models to assess U.S. emissions goal options (Fawcett et al., 2009). There are other modeling studies and projections that one could consider, but we found EMF22 to be particularly useful for a number of reasons:

- EMF22 relates global GHG concentration goals to global emissions reduction and to U.S. emissions reduction as well as reductions in other major countries (see details in subsequent sections of this chapter).
- EMF22 is a multimodel analysis, which helps mitigate concerns about results being skewed by the assumptions (e.g., about the key drivers such as population, economic growth, and energy use) built into any one model. The resulting spread of model results provides a relatively robust way to identify estimates for a corresponding domestic emissions budget and to clarify the level of potential uncertainty.
- Most previous multimodel evaluations of long-term climate goals (e.g., from the U.S. Climate Change Science Program [Clarke, 2007], earlier studies by the Energy Modeling Forum [de la Chesnaye and Weyant, 2006], and scenarios assessed by the IPCC [Fisher et al., 2007]) have assumed global, immediate action, and some even have a start date as far back as 2000. A

---

developing regions delay action by a few decades, then the 450 ppm CO_2-eq goal could be met only by the end of the century if concentrations are allowed to temporarily overshoot this goal.

U.S. EMISSION TARGETS

U.S. GHG emissions reductions will not by themselves have a decisive impact on global atmospheric GHG concentrations, but actions that the United States might take over the coming decades to reduce domestic emissions do need to be considered within the context of the ultimate goal of stabilizing global climate. To this end, the EMF22 U.S. study (Fawcett et al., 2009) calculated different cumulative U.S. GHG emission budget goals over the period 2012 to 2050, based on the selection of a base year and a corresponding emissions-reduction target to be achieved by 2050 (Table 2.1). They calculated that an 80 percent reduction from 1990 levels corresponds to a budget of

recent European study (the adaptation and mitigation strategies ADAM project (Edenhofer et al., 2010) evaluated stabilization targets with a special focus on the technical feasibility and economic viability of low stabilization scenarios but still with full global participation. In contrast, EMF22 explores international scenarios in which many developing countries delay emissions reductions, providing a broader context for understanding U.S. goals.

- EMF22 is the most recent large-scale, multimodel exercise exploring both U.S. and international emissions and concentrations goals. It therefore is based on the most recent understanding and historical data. In this way, it contains more recent assessments of underlying drivers and improvements to models used for analysis than previous work such as the reference scenarios from the IPCC's Special Report on Renewable Energy (Nakicenovic et al., 2000) or the stabilization scenarios in the U.S. Climate Change Science Program report on scenarios (Clarke, 2007).

- Many of the EMF22 models allow the examination of technology deployment requirements for meeting domestic emissions budgets. In Chapter 3, we test the feasibility of the proposed domestic emissions budget by comparing these deployment requirements against the technical potential of the relevant technologies to meet them.

We know of no other recent modeling exercise that exhibits all these useful features. Nevertheless, we also recognize some important limitations of the EMF22 study. These include, for instance, the question of how aerosols and land use impacts are represented in the models. We note in the subsequent sections where these limitations affect our analysis, but overall we do not believe these effects significantly change our main conclusions and recommendations.

167 gigatons (Gt) CO_2-eq[9] (and a 50 percent reduction corresponds to a budget of 203 Gt CO_2-eq[10]). Table 2.1 illustrates that the budget values do not change a great deal if different baseline years are selected. This is a useful characteristic of a cumulative budget, since the choice of baseline year is often an issue of contentious debate in policy negotiations.

We chose to round these EMF numbers and adopt a cumulative emissions budget in a range of 170 to 200 Gt CO_2-eq over the period 2012-2050 as a reasonably representative U.S. budget target, which can serve as a benchmark for developing policy recommendations and testing their feasibility. This budget range, which represents a significant change from business-as-usual U.S. emissions out to 2050, is roughly in line with the types of emissions-reduction goals found in many recent policy propos-

---

[9] A gigaton equals a billion (10^9) tons.

[10] These cumulative emissions levels are based on linear pathways from 2012 through 2050. Note, for comparison, that recent (2008) annual U.S. emissions were approximately 7 Gt.

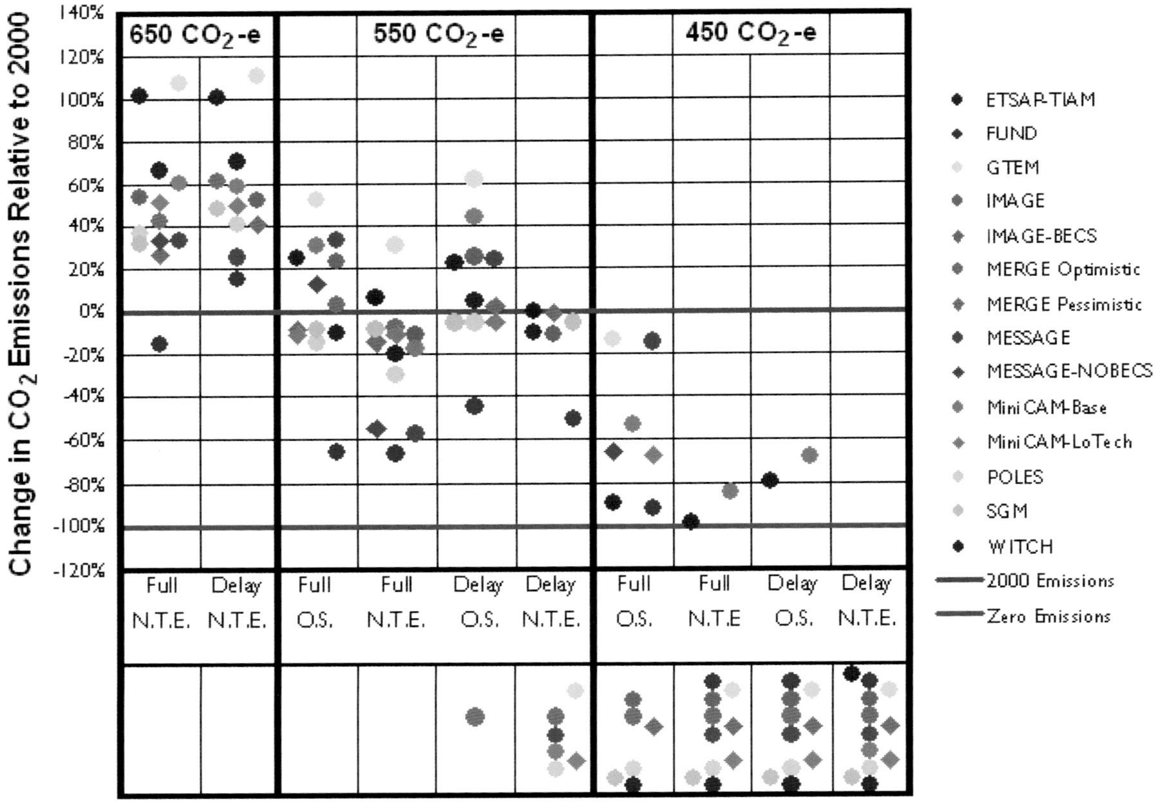

FIGURE 2.9 Global fossil and industrial CO_2 emissions in 2050 relative to those in 2000. Long-term concentration goals are 450, 550, or 650 CO_2-eq (Kyoto gases). The level of international action is either "Full," meaning all countries undertake emissions mitigation starting in 2012, or "Delay," meaning low- and middle-income countries do not begin climate mitigation until 2030 or beyond. The option to exceed or "overshoot" the long-term concentration goal this century was defined either as a not-to-exceed formulation (N.T.E. in the figure) or as an overshoot case (O.S.). The more stringent scenarios (starting with the 550 ppm CO_2-eq case, without immediate global participation), could not be represented in some or all models and are indicated as such at the bottom of the figure. SOURCE: Clarke et al. (2009).

TABLE 2.1 Cumulative U.S. GHG Emissions (Gt CO_2-eq) for the Period 2012 to 2050, Assuming Linear Emissions Reductions Throughout That Time Period, Beginning at a 2008 Emissions Level, and Assuming 100 Percent Coverage of GHG Emissions in the Economy

Base Year	Percent Below Base Year Emissions in 2050						
	83%	80%	65%	50%	35%	20%	0%
1990	164	*167*	185	*203*	221	239	262
2005	167	*171*	192	*213*	234	254	282
2008	168	*172*	194	*215*	237	258	287

SOURCES: Fawcett et al. (2009). Emissions data from 1990 and 2005 are from EPA (2009), and 2008 emissions projections are based on Paltsev et al. (2007).

als. See Figure 2.10 for a graphical representation of these goals. Note that these goals are presented as limits for cumulative U.S. domestic emissions (rather than goals to be met through international offsets). Excluding some sectors from emissions reductions, or allowing international and domestic offsets, could substantially alter the actual U.S. emissions reductions.

This budget range also relates usefully to the EMF22 international scenarios, where modeling results for global emissions reduction were disaggregated (using the global least-cost criteria described earlier) to indicate the degree of U.S. action associated with global goals under differing cases of international action (the same set of cases illustrated in Figure 2.9). The results show a considerable range of possible U.S. contributions, reflecting the large uncertainties inherent in any long-range (midcentury) projections. In general, however, the EMF22 results indicate that a 200 Gt CO_2-eq U.S. budget is roughly consistent with a long-term global goal of 550 ppm CO_2-eq, particularly if there is full international participation. This same U.S. budget could also be consistent with long-term global goals of 450 ppm CO_2-eq, but only if the option exists to overshoot the goal (with the attendant requirement for more aggressive actions beyond 2050) and if there are also immediate, aggressive, and comprehensive global GHG emissions-reduction efforts. The more stringent 170 Gt CO_2-eq U.S. budget is roughly consistent with the 550 ppm CO_2-eq global goals without overshoot or with delayed participation—or with the most idealized of the 450 ppm CO_2-eq goals (immediate, comprehensive international action, and the opportunity to overshoot the long-term goal prior to 2100). See Clarke et al. (2009) and Fawcett et al. (2009) for further details on these calculations.

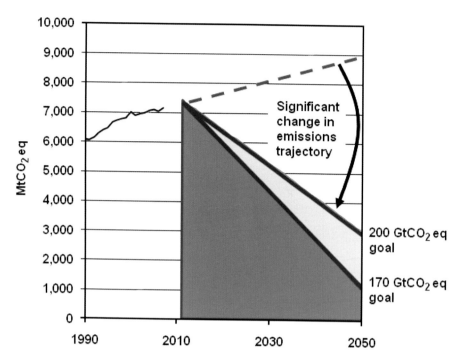

FIGURE 2.10 Illustration of the representative U.S. cumulative GHG emissions budget targets: 170 and 200 Gt CO_2-eq (for Kyoto gases) (Gt, gigatons, or billion tons; Mt, megatons, or million tons). The exact value of the reference budget is uncertain, but nonetheless illustrates a clear need for a major departure from business as usual.

As mentioned above, these emissions budgets are for gross emissions in the United States and do not include sources and sinks from land use, land-use change, and forestry (LUCF). If LUCF emissions are net positive (emissions), it will make attaining these budgets more difficult. If LUCF emissions are net negative (sinks), which is the current trend, it will make attaining these budgets easier.

There are many differing views on the relative burdens that different countries should bear to address climate change. For example, it has been suggested that the United States should make more stringent emissions-reduction efforts, based, for instance, on precautionary concerns that lower global concentration targets are needed, or based on "fairness" arguments that high-income countries, having produced most of the GHG emissions to date, should shoulder a larger share of future emission reductions. For example, the German Advisory Council on Global Change (2009) developed a global emissions budget and applied the criterion of equal per capita emissions among all countries; this calculation allocates to the United States a budget of 35 Gt CO_2-eq over the same time period of the EMF22 budget.

Conversely, actions by other countries influence the U.S. contribution required to meet any global concentration goal. The EMF study indicates that, in general, the effect of delaying action is not to dramatically alter global emissions reductions required for 2050 but rather to cause a shift in the distribution of emissions among regions. Thus, emissions reductions not undertaken by one large country (or group of smaller countries) must be made up for by other countries in order to achieve atmospheric GHG stabilization at the levels explored in the scenarios above.

If the United States elected to make additional commitments beyond the least-cost allocated domestic budget mentioned above, it would be reasonable to achieve these additional commitments through a mechanism for investing in emissions reductions elsewhere in a way that does not increase the total cost of meeting the global emissions budget. The purchase of international offsets or participation in a global carbon pricing system would in principle provide such a mechanism. As discussed in Chapter 4, however, these mechanisms must be designed to ensure the emissions reductions are "real, additional, quantifiable, verifiable, transparent, and enforceable" and do not result in emissions leakage, all of which are difficult challenges to address.

If the United States sought to lessen its domestic emissions-reduction requirements by purchasing offsets from other countries, then to avoid double counting, the countries selling offsets could not take credit for these emissions reductions if they establish their own national emissions budget. Economic theory suggests that the solution to this kind of problem is to compensate the seller for the loss of the purchased emissions reduction. For example, the United States might purchase an offset from Country A for the price of the offset itself plus the future cost that Country A may face in reducing its emissions through actions that are more costly than the original offset. It is beyond our scope to recommend how to design an international offset system that addresses this issue, but it is worth noting that a system without this form of additional compensation may be resisted by countries interested in selling offsets.

Finally, we note that, because of the uncertainties and judgment involved, the initial U.S. domestic emissions budget may not be stable over time. Indeed, changes to the emissions budget—up or down—are to be expected. This is an unavoidable problem with a scientifically complex, politically controversial, and long-term problem like climate change. This fact is a primary reason for ensuring that the U.S. policy framework for limiting GHG emissions is both durable and adaptive, as discussed in Chapter 8.

IMPLICATIONS OF U.S. EMISSION GOALS

To evaluate the implications of the representative U.S. emissions budgets discussed above, we have drawn from the U.S. component of the EMF22 study[11] (Fawcett et al., 2009). These analyses help illustrate one of the reasons why cumulative emissions goals are an effective way to set long-term targets: They allow flexibility in emissions over time. Emitters might either bank permits (by reducing more than the target in a particular year) or borrow permits (so that they can emit more than they are allotted in a given year). The opportunity to bank emissions rights for future use, or to borrow emissions rights from the future, means that the actual emissions pathway to any cumulative 2050 goal will probably not be linear (see Fawcett et al. [2009] for discussion of the forces that could influence the degree to which emitters choose to bank or borrow emissions rights).

Figure 2.11 shows estimates from the same study for CO_2 emissions reduction from electric power generation and transportation. For the 203 and 167 Gt CO_2-eq scenarios, electricity sector emissions in 2050 are reduced by an average of approximately 90 and 100 percent, respectively; in the transportation sector, emissions fall by an average of approximately 20 and 30 percent, respectively. Overall, the electricity sector reduces emissions to levels well below the target while transportation-sector emissions remain well above the target. This reflects differences in the emissions-reduction options across sectors. These projections are of course influenced by assumptions built into the models, which are generally based on technologies and processes that we know how to characterize. It is possible that unknown technological breakthroughs or major socioeconomic shifts could lead to a very different picture in the future.

Ideally, the costs of U.S. emissions reductions are measured as a change in the well-being of Americans. However, all measures that aggregate across the entire population give rise to ethical questions about how to weigh differences in the well-being of separate individuals and groups within that population. Hence, for practical purposes, aggregate economic metrics are often used to represent the change in well-being that is brought on by emissions reductions. Typical metrics include CO_2 prices (which, while not a direct measure of cost, can be used to determine effects on basic energy goods such as gasoline or natural gas for home heating and cooking) and overall changes to economic output such as GDP.

Figure 2.12 shows the carbon prices estimated in the U.S. component of the EMF22

---

[11] The U.S. component of the EMF22 study included six models that explored specific domestic goals (see Fawcett et al., 2009). Note that this study used budget targets of 167 to 203 Gt CO_2-eq, rather than our "rounded" targets of 170 to 200 Gt CO_2-eq.

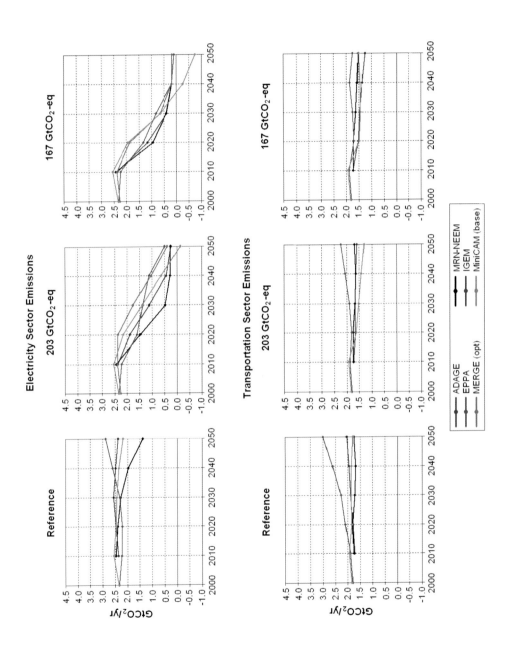

FIGURE 2.11 EMF22 electricity and transportation CO_2 emissions and reduction for 167 and 203 Gt CO_2-eq goals. Note that the electricity sector generates greater emissions reductions than the transportation sector in all scenarios. SOURCE: Fawcett et al. (2009).

study for the cumulative emissions scenarios discussed above. Several insights emerge: First, there is a distinct range of prices. For example, under the 167 Gt CO_2-eq goal, the 2020 carbon price ranges from roughly $50 to $120 per ton of CO_2. The differences among these estimates stem largely from differing expectations about the technologies that will be available and the ability to deploy these technologies effectively. Note that this includes not just energy supply technologies but also technologies to reduce emissions in end-use and industrial applications. Second, the effect of CO_2 prices on energy gives a sense of the economic burden that would be imposed by emissions-reduction policies. Ultimately the effect of CO_2 prices is to increase the cost of carbon-intensive energy and of products that use energy as an input. Table 2.2 shows the effect of a $100 per ton CO_2 price on the costs of key fuels.

In all the scenarios and for all the models, carbon prices exhibit a steady increase. This is a key feature found in virtually all emissions-reduction studies. Because the stringency of the reductions must increase over time as emissions are eventually driven toward zero, the costs must go up over time. Thus, meeting the sorts of cumulative goals proposed here will require an increasing commitment with an increasing cost. Note, however, that increasing prices are based on the assumption that the exact degree and nature of future improvements in important drivers such as technology, economic growth, and population growth are known with certainty. If, for example, technology were to advance substantially more rapidly than expectations (for example, if there were to be a radical technological breakthrough), the CO_2 price would rise less aggressively. Conversely, less-than-expected technological advance could drive price increases even higher. (See Chapter 5 for more discussion about technological innovation as a key factor for modulating GHG emission-control costs.)

Aggregate economic indicators such as GDP or consumption losses are another common way to represent the costs of GHG emissions reduction. Given the simplifications required for a model to represent the national economy, these sorts of estimates are best viewed as informative in relative terms but highly uncertain in absolute terms. Figure 2.13 shows projected U.S. GDP under reference cases and the two budget scenarios, looking across the different models used in the EMF22 study. There is a large degree of uncertainty in future economic growth, for instance, with reference projections for 2030 varying by about 22 percent from the highest to lowest estimated values.

An important insight emerges when comparing projected economic growth in a "no-policy" case (i.e., reference scenario) to a "policy" case (i.e., with mandates for the budget targets discussed earlier); that is, although climate action does put downward pressure on economic growth, the effects over the next several decades are generally

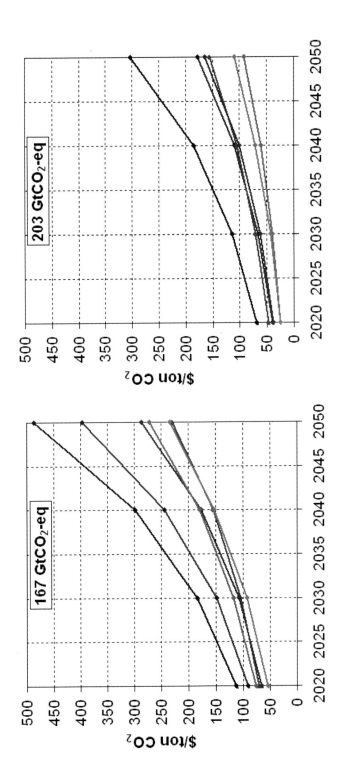

FIGURE 2.12 Carbon prices across EMF22 scenarios for 167 and 203 Gt CO_2-eq goals. In all scenarios, carbon prices increase over time, and prices are higher for the more stringent (167 Gt CO_2-eq) emissions budget. SOURCE: Fawcett et al. (2009).

TABLE 2.2 Effect of Carbon Prices on Energy Prices

Carbon Price/Fuel Price and Type	Current 2009 Prices	Cost of Carbon Content	Total End-User Price	Added Cost (%)
Metric ton of CO_2		$100.00		
Metric ton of Carbon		$366.67		
Crude oil ($/bbl)	$65.27	$42.80	$108.07	66
Regular gasoline avg. ($/gal)	$2.32	$0.88	$3.20	38
Utility coal avg. ($/short ton)	$46.34	$221.05	$267.39	477
Residential natural gas avg. ($/tCf)	$12.47	$5.44	$17.91	44

NOTE: The additional costs are based on a $100 per ton CO_2 price. Percentage added cost is dependent on base costs at any point in time and therefore subject to some variability.
SOURCE: EIA December 2009 *Monthly Energy Review*.

modest in comparison to the degree of growth. For instance, in the most optimistic projection of the EMF22 study, for the reference case, economic growth for the period 2010 to 2130 increases by 88 percent. When the most stringent emissions-reduction targets are imposed, economic growth still increases by 83 percent. In the most pessimistic projection, economic growth is 52 percent in the reference case; this is changed to 51 percent for the 167 Gt CO_2-eq target case. Economic losses increase over time, so the expectation is that they will be larger through and beyond 2050 than they are through 2030. It is important to note that none of these GDP impacts include estimates of the welfare benefits that would be associated with reducing GHG emissions. Note also that these studies assume efficient national policy architectures resembling an economy-wide cap-and-trade system or carbon tax. Less efficient approaches could substantially increase costs.

Figure 2.14 presents another way of evaluating differences across models: to compare the impacts as a percentage of reference GDP. By 2030, GDP losses range from 0.5 to 4.5 percent. The range of estimates highlights the tremendous uncertainty surrounding the costs of climate action, which is due to differences among analyses including different economic growth and GHG emission levels in the reference (no-policy) case, differences in how households respond to higher energy prices, and variations in the deployment and effectiveness of mitigation technologies. This range is roughly consistent with previous studies, such as the U.S. CCSP scenarios (Clarke, 2007). In no scenario does growth stop or does the economy decline; rather, in all cases, the effect of the emissions-reduction policy is to delay the achievement of higher GDP levels. As

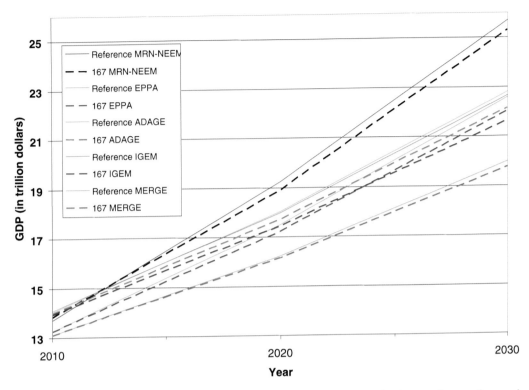

FIGURE 2.13 Projected U.S. GDP under reference cases and a 167 Gt CO_2-eq budget goal across five models used in the EMF22 study. SOURCE: F. de la Chesnaye, EPRI.

noted above, none of these GDP impacts includes estimates of the benefits that would be associated with reducing GHG emissions.

A Congressional Budget Office report on the economic effects of GHG limiting policies (CBO, 2009) arrives at the same finding: that there is an upfront cost to the economy, but it will be relatively modest. The main impact of GHG limiting policies would be on energy expenditures, which accounted for about 9 percent of GDP in 2006 (EIA, 2009). A resulting price on GHG emissions ($/tCO_2$-eq) would lead to higher delivered-energy prices which in turn would lead to decreased economic output. In general, the economy will shift production, investment, and employment away from sectors related to the production of carbon-based energy and energy-intensive goods and services and toward sectors related to the production of alternative energy sources and non-energy-intensive goods and services.

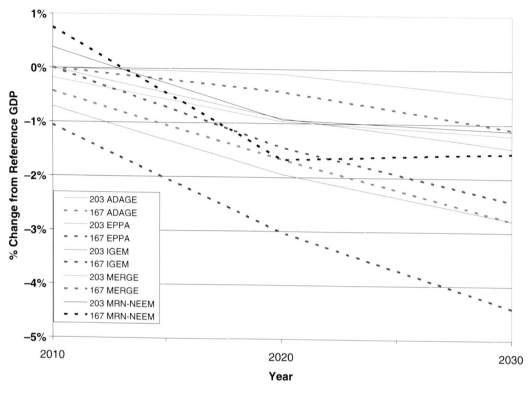

FIGURE 2.14 Impact of 167 and 203 Gt CO_2-eq budget targets as a percent of reference GDP across five models used in the EMF22 study. Negative GDP losses (projected increases) in the near term are due to households increasing expenditures in the near term, in expectation of higher prices in the future. SOURCE: F. de la Chesnaye, EPRI.

KEY CONCLUSIONS AND RECOMMENDATIONS

Future U.S. and global GHG emissions will be driven by trends in population, economic activity, intensity of energy use, and technological developments. Thus, long-term GHG emissions trends are difficult to predict with certainty. It is highly likely, however, that emissions will continue to rise in the coming decades without concerted new emissions-reduction policies.

Recent integrated assessment modeling studies indicate that limiting the increase in global atmospheric GHG concentrations to 450 ppm CO_2-eq this century (which can be related, in probabilistic terms, to the goal of limiting global mean temperature rise to 2°C above preindustrial levels) would require aggressive emissions-reduction efforts by all major GHG-emitting nations, starting within the next few years.

From a quantitative perspective, significant U.S. emissions reductions will not by themselves substantially alter the rate of climate change. Although the United States has the largest share of historic contributions to global GHG concentrations, this relative share will decrease over time. All major economies will need to reduce emissions substantially in concert with the United States.

Although long-term global mean temperature change and global atmospheric GHG concentrations are essential outcomes for policies to limit future climate change, they are not sufficient metrics for setting a domestic policy goal. The domestic goal will need to be one that policy can affect directly and for which progress can be measured directly. We recommend that the U.S. goal be framed as a cumulative emissions budget over a set period of time.

Identifying global temperature and atmospheric GHG concentration targets, and linking these to global and U.S. emissions-reduction goals, involves numerous scientific uncertainties as well as ethical and political judgments. We thus do not attempt to recommend definitive U.S. emissions-reduction goals here. However, as a benchmark for the analyses in this study, we conclude that a reasonable range for representative budget goals is 170 to 200 Gt CO_2-eq for the period 2012 to 2050. These numbers were chosen because they roughly correspond to the goals of reducing U.S. emissions by 80 and 50 percent, respectively, by 2050—targets that have been used in many recent policy proposals—and because studies indicate that they are roughly consistent with the goals of limiting global GHG concentrations to 450 and 550 ppm, respectively (using global least-cost criteria for allocating a global emissions budget).

This representative U.S. emissions budget range is suggested as actual reductions in domestic emissions rather than a goal to be met through international offsets. A commitment to deeper emissions reductions, as some suggest is warranted on precautionary or fairness grounds, could possibly be achieved through mechanisms for investing in emissions reductions internationally, including the purchase of international offsets if they are truly additional and verifiable (which is discussed further in Chapter 4).

The costs of meeting these emissions-reduction goals are highly uncertain and depend heavily on the available technological options. Recent research estimates the prices of CO_2 per ton that would result from the emissions budget scenarios mentioned above; across these scenarios, however, climate action reduces U.S. GDP by between 0.5 and 4.5 percent in 2030. In all the scenarios, however, GDP continues to grow substantially through midcentury. None of these GDP impacts include estimates of the benefits that would be associated with reducing GHG emissions. Also note that these studies assume well-constructed, efficient national mitigation policies. Less efficient approaches could substantially increase the costs of mitigation.

CHAPTER THREE

Opportunities for Limiting Future Climate Change

I n Chapter 2 we recommended that the United States adopt a budget for cumulative greenhouse gas (GHG) emissions, and in this chapter we evaluate the opportunities and challenges involved in meeting this budget. To make this evaluation, we first examine a wide array of opportunities for reducing CO_2 emissions from U.S. energy consumption (summarizing only briefly the topics that are addressed in detail in the National Research Council [NRC] study *America's Energy Future*[1]), as well as enhancing CO_2 sequestration and reducing emissions of other GHGs. We then examine whether aggressively exploiting near-term emissions-reduction strategies (using technologies available now or in the near future) can yield the kinds of emissions reductions needed to achieve the budget goals. Gauging the prospects of technology in this way, even in a very general sense, is essential for understanding the urgency and nature of policy actions needed. Finally, we consider how the technological advancements highlighted in this chapter fit into a larger set of research questions about the interplay of technology with social and behavioral dynamics. Chapter 4 examines the policy approaches needed to exploit the emissions-reduction opportunities highlighted here.

OPPORTUNITIES FOR LIMITING GHG EMISSIONS

CO_2 emissions from fossil fuel combustion in the energy system account for approximately 82 percent of total U.S. GHG emissions. The amount of fossil fuels consumed is driven by economic, population, and demographic factors that affect overall demand for goods and services that require energy to produce or deliver; by the efficiency with which the energy is used to provide these goods and services; and by the extent to which that energy comes from fossil fuels (i.e., the carbon intensity of energy supplied). Opportunities exist in each of these areas to reduce CO_2 emissions. In addi-

---

[1] *America's Energy Future* consisted of three panel reports: (1) *Electricity from Renewable Resources: Status, Prospects, and Impediments*, (2) *Liquid Transportation Fuels from Coal and Biomass*, and (3) *Real Prospects for Energy Efficiency in the United States*, and an overarching report, *America's Energy Future: Technology and Transformation* (NRC, 2009a). More information is available at *http://sites.nationalacademies.org/Energy/*, and all reports are available at *http://www.nap.edu*.

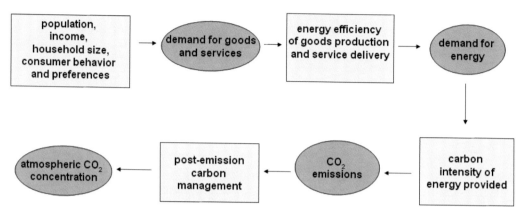

FIGURE 3.1 The chain of factors that determine how much CO_2 accumulates in the atmosphere. Each of the boxes represents a potential intervention point. The blue boxes represent factors that can potentially be influenced to affect the outcomes in the purple circles.

tion, atmospheric CO_2 concentrations can be altered by managing carbon sinks and sources in the biosphere and by chemical means that withdraw CO_2 from the atmosphere (post-emission carbon management). Figure 3.1 summarizes these major areas of opportunity, or potential points of intervention, in the effort to reduce atmospheric GHG concentrations. Each of these intervention points is discussed individually in the following sections, but it should be acknowledged that, in some instances (for example, in shaping future urban development patterns), major advances will require systems-level solutions, involving intervention at several of these points simultaneously.

Influencing Demand for Goods and Services that Require Energy

The first box in Figure 3.1 identifies various factors that have been shown to influence the overall level of demand for goods and services in an economy. Curbing U.S. population growth (either through policies to influence reproductive choices or immigration), or deliberately curbing U.S. economic growth, almost certainly would reduce energy demand and GHG emissions. Because of considerations of practical acceptability, however, this report does not attempt to examine strategies for manipulating either of these factors expressly for the purpose of influencing GHG emissions.

An issue of key relevance is the practicality and acceptability of intervening to alter consumer behavior and preferences in ways that would reduce the demand for goods and services that result in energy consumption and GHG emissions. (We note this is different from the question explored in the following section: how to meet demand for goods and services in a way that uses less energy and/or emits fewer GHGs per unit of output.) What is the potential for changing consumer behavior and preferences? The United States has larger per capita energy use than many other countries with an equal or higher standard of living, such as Japan and most European countries. This differential is no doubt due to a variety of economic, demographic, geographic, and cultural factors, including differences in energy prices and energy efficiency. The extent to which the gap derives from differences in consumer desires for energy-intensive goods and service is less clear.

Consumer choices among market offerings in different societies shape demand for everything from living space and electric appliances to dietary choices. For instance, the social dynamics leading to larger, more dispersed dwellings, manifest in suburban development, is an important factor in contemporary U.S. energy use. The pattern of low-density suburban development gained momentum in the 19th century with the advent of electric street cars, and it accelerated during the mid-20th century after the widespread introduction of automobiles and freeways lowered the cost of living and working farther from city centers. Social preferences for lower density and more living space thus have deep roots in American society, and changing these patterns can be extremely challenging. Yet many of America's central cities and inner-ring suburbs have remained vital over the past century; many urban planners and advocates for "smart growth" find that interest in denser development is growing. For instance, as the population ages, many older people seek smaller homes closer to amenities and services (Myers and Gearin, 2001). Immigrant groups have also tended to migrate to central cities and inner suburbs. Technologies that lower the cost of living in denser communities (for instance, quality, affordable transit, and car-sharing programs) have been proposed as an impetus for more compact living and working environments (Sperling and Gordon, 2009). Box 3.1 summarizes key findings from a recent NRC study that evaluated the linkages among urban development patterns and GHG emissions in depth.

Environmental awareness about energy security and global climate change are on the rise (Curry et al., 2007). Levels of concern fluctuate with the changing importance of other social, economic, and environmental issues, and the strength of concern varies across segments of the population (Leiserowitz et al., 2008). But it remains clear that much of the U.S. population views climate change as an important public policy problem (Pew Center, 2009a). As Americans become increasingly informed about climate change, does this concern translate into new consumption patterns?

Social science research in this area suggests that information and attitudes alone are unlikely to prompt the sorts of changes in long-standing patterns of technology use

BOX 3.1
Urban Development and Transportation Energy Demand

Sprawling, automobile-dependent development patterns are a major factor underlying U.S. dependence on petroleum and thus much of our GHG emissions. There is growing interest in the idea that more compact, mixed-use development will reduce vehicle miles traveled (VMT), make alternative modes of travel more feasible, and thus offer an important strategy for reducing CO_2 emissions. The NRC Transportation Research Board (TRB) recently examined this question of whether petroleum use and GHG emissions could be reduced by changes in development patterns. Below is a brief overview of some key findings (from NRC, 2009d).

Developing at higher population and employment densities means trip lengths will be shorter on average, walking and bicycling can be more competitive alternatives to the automobile, and it is easier to support transit. Increasing density alone, however, is generally not sufficient to reduce VMT by a significant amount. A diversity of land uses that result in desired destinations (e.g., jobs, shopping) being located near housing, and improved accessibility to these destinations, are also necessary. Development designs and a street network that provides good connectivity between locations and accommodates nonvehicular travel are important. Finally, demand management policies such as lowering parking requirements and introducing market-based parking fees are also needed. The effects of compact development will differ depending on where it takes place: Increasing density in established inner suburbs and urban core areas is likely to produce substantially more VMT reduction than developing more densely at the urban fringe.

The TRB committee developed illustrative scenarios to estimate the potential effects of more compact, mixed-use development on reductions in energy consumption and CO_2 emissions. An "upper bound" scenario (with 75 percent of new housing units steered into more compact development and residents of compact communities driving 25 percent less) could lead to reduced VMT and associated fuel use and CO_2 emissions by about 7 to 8 percent less than the base case by 2030, and 8 to 11 percent less by 2050. A more moderate scenario (with 25 percent of new housing units built in more compact development and residents of those developments driving 12 percent less) could lead to reductions in fuel use and CO_2 emissions of about 1 percent by 2030, and 1.3 to 1.7 percent by 2050. Overall then "the committee believes that reductions in VMT, energy use, and CO_2 emissions resulting from compact, mixed use development would be in the range of ~1 to 11 percent by 2050, although the committee members disagreed about whether the changes in development patterns and public policies necessary to achieve the high end of these findings are plausible."

It is important to keep in mind, however, that these potential emissions reductions resulting from land-use changes would be occurring in the context of an overall increasing baseline of VMT; thus, even at the high end of the optimistic scenario, VMT in 2050 may be higher than it is today.

and consumption of energy-intensive goods that are required for making significant reductions in GHG emissions. For example, fostering significant progress in residential energy conservation requires not only changes in public awareness and concern but also changes in market product offerings and changes in the behavior of both producers (home builders) and consumers (home buyers or renters) (Lutzenhiser et al., 2009; Stern, 2008). Long-term sustained changes will be driven by the interactions of technology markets, the policy environment, and consumer choices.

Changes in the demand for goods and services are an expected and desired outcome of carbon pricing strategies—especially through the substitution of more energy-efficient goods and production processes. However, until such a system is enacted at broad scales, it remains unclear just how much consumer behavior and overall demand for goods and services can be modified through prices alone. Consumer responses to financial incentives in the past have been highly variable. The largest impacts are seen in cases where complementary policies and nonfinancial incentives have also been provided (Gardner and Stern, 2002; Stern, 1986).

Public interest supported by thoughtful policy and good communication has been shown to be an effective combination for changing consumer behavior (NRC, 2005). The long-term successes from sustained public health information campaigns, coupled with disincentives and penalties (e.g., in the cases of smoking and drunk driving), suggest that public attitudes can be modified over time in ways that significantly affect behavior and demand. Public policies devised to reinforce changes that are already occurring in public attitudes and consumer preferences are likely to be more effective in bringing about the changes needed to dramatically lower GHG emissions.

For instance, many processes involved in water consumption require a significant amount of energy. The energy used to pump and purify water, make hot water, and treat wastewater is a large driver of electricity use for many municipal governments. Sensitivity to water use in some parts of the country is already high as a result of past droughts and increasing water scarcity. In these places, state and local governments have developed significant expertise in building effective communication and policy strategies. Publicity campaigns and other actions to promote water conservation may thus be an area where public policy could contribute to reducing energy demand. The same is true for residential energy conservation, where some states and locales have long-standing commitments and significant expertise in interventions that combine technology and behavior change to reduce demand for electricity and natural gas— although considerable work remains to be done in this area (Lutzenhiser et al., 2009).

Improving the Efficiency of Energy Use

Many opportunities exist to improve the efficiency of energy use. Total U.S. energy consumption today is 40 percent higher than it was in 1975. At the same time, energy intensity, measured as energy use per dollar of gross domestic product, has steadily fallen, averaging a decline of 2.1 percent per year (NRC, 2009a). About 70 percent of the decline in energy intensity is estimated to have resulted from improvements in energy efficiency (IEA, 2004). If current trends continue, U.S. energy intensity would drop by 36 percent over the next two decades. Despite these impressive gains, however, almost all other developed nations continue to use significantly less energy per capita than the United States (NRC, 2009a).

What is the potential for further gains in U.S. energy efficiency? Most analysts believe that the technical potential in the aggregate is large and much of it can be realized, especially if the price of energy increases. Judgments about the technical potential for energy-saving technologies and practices and their deployment to common use, however, are often fraught with uncertainty. Indeed, it has long been perplexing why consumers and businesses do not take greater advantage of what seem to be cost-effective energy-efficiency opportunities, that is, why they do not choose technologies that appear to quickly "pay for themselves" in energy cost savings. A new technology that appears promising may encounter various market barriers that hinder its implementation, or the technology itself may be lacking some important attribute (e.g., in reliability, durability, function) that makes it less cost-effective than expected. Regardless, consumers may be slow to adopt a new technology because of uncertainty about real-world savings potentials, future energy prices, and the prospects of even better technologies coming to market in the future.

A host of market and institutional barriers have been identified in the literature (Brown et al., 2007; DOE, 2009). For example, in the "principal/agent problem," those paying for the technology and those benefiting from it are not the same. This barrier is significant and widespread in many energy end-use markets (Prindle, 2007). The landlord-tenant relationship is the classic example: If a landlord buys the energy-using appliance while the tenants pay the energy bills, the landlord is not motivated to invest in efficiency. Often monthly energy costs are included in the rent, providing the tenant with no incentive to conserve. About 90 percent of all households in multifamily buildings are renters, which makes this a major obstacle to energy efficiency in urban housing markets. Conflicting landlord and tenant motivations are also a problem with commercial buildings, many of which are rented or leased. In addition, many buildings are occupied by a succession of temporary owners or renters, each unmotivated to make long-term improvements that would mostly reward subsequent occupants (Brown and Southworth, 2008).

Other obstacles can stem from the lack of basic information, such as consumers not knowing how a particular appliance may be affecting their monthly household electricity bill. Moreover, given the early stage of deployment (at least in the United States) of many energy-efficient technologies (e.g., cogeneration, light-emitting diodes, and plug-in hybrid electric vehicles), obtaining reliable information can be costly, time-consuming, and perhaps not possible (Worrell and Biermans, 2005). Such market barriers have long been used to justify public policies devoted to boosting energy efficiency, prominent examples being the Corporate Average Fuel Economy (CAFE) program, Energy Star, appliance and vehicle labeling requirements, and building energy codes (DOE, 2009).

Some barriers to adoption of energy-efficient technologies result from government regulations, subsidies, and penalties that were designed to address goals in areas other than energy. Risk aversion often limits the variety of technologies offered in the market. Constraints can also be imposed through various community standards and practices, such as homeowner association rules that require the use of particular materials and design elements or that prohibit others (e.g., white roofs, clotheslines, and shade trees).

The NRC study *America's Energy Future* (AEF) (NRC, 2009a,b) included a comprehensive review of energy-efficient technologies and processes in the sectors of industry, residential and commercial buildings, and transportation. The goal was to identify energy-saving technologies and practices that are currently ready for implementation, that need further development, or that exist just as concepts but are sufficiently promising to offer major efficiency improvements in the future.

Overall, AEF estimates that the potential cost-effective energy savings range (from the conservative to the optimistic) from 18.6 to 22.1 quads[2] in 2020 and from 30.5 to 35.8 quads in 2030. Comparing this to the Energy Information Administration (EIA, 2009) forecast for "business as usual" consumption (105.4 quads in 2020 and 113.6 quads in 2030), this means a potential for savings of 18 to 21 percent in 2020 and 27 to 32 percent in 2030. This more than offsets the EIA's projected increases in energy consumption through 2030, but it still falls short of achieving the very large GHG emissions reductions needed overall.

---

[2] A *quad* is a unit of energy equal to 1.055×10^{18} joules (1.055 exajoules or EJ). It is a unit commonly used in discussing global and national energy budgets.

Many studies, including the AEF assessment, examine efficiency opportunities by energy-use sector, such as transportation, industry, and buildings, and so we follow this construct below. Another useful vantage point, however, is to focus on the perspective of the actors actually making investment and purchase decisions, such as the household-level actions discussed in Box 3.2.

BOX 3.2
Household-Level Actions to Increase Conservation and Energy Efficiency

It has been estimated that households contribute roughly 40 percent of national GHG emissions through direct energy use in homes and nonbusiness travel, plus an additional 25 percent indirectly through GHGs emitted in the production, distribution, and disposal of consumer goods and services (Bin and Dowlatabadi, 2005; Gardner and Stern, 2008). There are a variety of ways in which household-level actions can enhance energy conservation and efficiency. Analyses find that the greatest potential to lower direct household energy use occurs in two main areas: (1) the choice of more energy-efficient motor vehicles and (2) home space conditioning technology (insulation, windows, furnaces, and air conditioners). Gardner and Stern (2008) estimate that these two types of efficiency improvement can save nearly 20 percent of total household energy consumption each, for an average household that has not already undertaken the action.

Dietz at al. (2009) found that, aggregated across all U.S. households, the technical potential for emissions reduction is approximately 9 percent for phasing in more fuel-efficient vehicles, 6 percent for home weatherization and adoption of more efficient space conditioning equipment, 5 percent for more efficient household appliances, and 5 percent for universal adoption of compact fluorescent lighting. The study suggests that, with effective incentive programs, the great bulk of these efficiency improvements could realistically be achieved.

In addition, Dietz et al. found emissions-reduction opportunities (of ~17 percent of household direct emissions) resulting from changes in the maintenance and use of household equipment. However, they note that some of these additional changes—such as carpooling—are often resisted because they are seen as sacrificing time, comfort, or convenience. Policies designed to achieve optimal short-term emissions reductions will need to take into account the different opportunities and constraints associated with different kinds of behavioral change.

A recent Department of Energy (DOE) effort examining federal policies to reduce CO_2 emissions in the residential sector (based on current knowledge of behavioral barriers) found that greater understanding of household behavior is needed to optimize the design of such policies (Brown et al., 2009a).

Building-Sector Efficiency

The AEF committee concluded that the buildings sector (including both private housing and commercial buildings) offers the greatest potential for energy savings from
efficiency gains. Most energy use in the building sector is in the form of electricity,
followed by natural gas. There are numerous options to reduce this energy use, ranging from simple insulation and caulking to highly sophisticated appliances (Granade,
2009).

Take lighting as an example: Solid-state lighting is an important emerging technology with significant energy-savings potential. Compact fluorescent lights are a major
improvement over incandescent lamps with respect to efficiency, but they have disadvantages in other respects (e.g., they contain mercury, are difficult to dim, are not a
point light source, and are not "instant on"). Light-emitting diodes (LEDs) do not suffer
from these disadvantages, and the best LEDs are now more efficient than fluorescent
lamps (Craford, 2008). DOE (2006a) projects that LEDs will yield a 33 percent savings by
2027, relative to projected lighting energy use without LEDs. Since lighting accounts
for about 18 percent of primary energy use in buildings, the savings from this one
technology alone could amount to 6 percent of energy use in buildings by 2027.

Examples of other promising technologies (which can be applied to the existing building stock as well as new construction) are reflective roof products, advanced window
coating, natural ventilation, and smart heating and air-conditioning control systems.
Technologies available to reduce consumption in water heating (the second largest
consumer of energy in homes) include alternative heat pump water heaters, water
heating dehumidifiers, solar water heaters, and tankless water heaters (Brown et al.,
2007). The efficiency of cooling buildings can also be aided by design strategies such
as planting shade trees and replacing blacktop roofs with light-colored materials that
reflect away more sunlight and drastically reduce heat absorption.

Collectively, existing technology opportunities for residential buildings could save
over 500 terawatt hours (TWh) per year, more than one-third of the electricity now
used in residences and about twice the growth expected by 2030 (EIA, 2009).[3] The
commercial sector should be able to show even greater savings, about 700 TWh
(Brown et al., 2008).

---

[3] It should be noted that new homes comprise roughly 1 percent of the housing stock in any given year,
leaving much of the opportunity for energy and GHG reductions to the rehabilitation of existing homes and
disclosures at the time of their sale or lease.

Industrial-Sector Efficiency

There are numerous examples of how advanced sensors, intelligent feedback, and continuous process controls can offer industry-wide energy-savings potential. For example:

- In the papermaking industry, fiber optic and laser sensors can monitor water content, sheer strength, and bending stiffness of paper, both saving energy and improving paper quality (see *http://www.physorg.com/news4221.html*).
- Blending fly ash, steel slag, and other recycled materials with cement could cut energy consumption in the cement industry by 20 percent (Worrell and Galitsky, 2004).
- Data indicate that most U.S. petroleum refineries can economically improve distillation efficiency by 10 to 20 percent with improved systems such as gas separation technologies, corrosion-resistant metal- and ceramic-lined reactors, and sophisticated process control hardware and software (DOE, 2006b; Galitsky et al., 2005).
- Motors, the largest single category of electricity end use in the U.S. economy, offer considerable opportunity for electricity savings through technology upgrades and system efficiency improvements (achieved by selecting the appropriately sized and most efficient available motor for the application at hand). Next-generation motor and drive improvements, including the use of superconducting materials, are currently under development (NRC, 2009a).

The AEF committee pointed out that many of these approaches provide multiple ancillary benefits such as improved productivity, product enhancements, and lower production costs. They recognized, however, that risk aversion and uncertainty over future prices for electricity and fuels can lead many firms to defer decisions on energy-efficiency investments. The concern with such deferrals is that, once an asset is installed, it locks in a fixed level of energy efficiency for years or even decades (IEA, 2008). This adds to the importance of aggressively pursuing "windows of opportunity" to put efficient technologies and systems in place. NRC (2009b) estimates that investments in available efficiency technologies (including growth in combined heat and power production) could reduce energy consumption in the industrial sector by 14 to 22 percent (about 4.9 to 7.7 quads) over the next decade.

Transportation-Sector Efficiency

Concerns over U.S. dependence on imported oil provide an additional motivation for increasing energy efficiency (see Chapter 6). Cars and light trucks are the main source of energy consumption in the U.S. transportation sector, accounting for ~65 percent of fuel use. Improving automobile fuel economy has thus been a central focus of federal energy policies dating back more than 30 years to the establishment of the Corporate Average Fuel Economy (CAFE) Program. Most recently, the 2007 Energy Independence and Security Act (EISA) mandates substantial increases in the CAFE fuel economy standards for new cars and light trucks sold over the next decade, requiring a combined 35 miles per gallon for vehicles sold in 2020 (representing a 40 percent increase from today). In September 2009, the Obama Administration proposed GHG emissions performance standards for new cars and light trucks that are intended to accelerate these fuel economy gains.

The AEF report concluded that these increases in vehicle fuel economy will be difficult but possible to meet, since many technologies are available that could be implemented at relatively modest cost. Some are already in use and could be expanded rapidly over the next decade (e.g., cylinder deactivation, direct injection, diesel engines, and hybrid electric vehicles). Others, such as plug-in hybrid electric vehicles, have the potential to start penetrating the market during the next decade, possibly leading to all-electric battery vehicles. The report stresses, however, that there is no assurance these improvements will be introduced on a wide scale in this time frame, especially if motor fuel prices do not rise and create incentives for consumers to demand more energy-efficient vehicles.

The question of whether more stringent fuel-efficiency standards would be warranted at a later date depends upon how the policy environment and technological capabilities evolve in the next two decades, as discussed further in Chapter 4. Overall, NRC (2009a) estimates an approximate energy-savings potential for light-duty vehicles of 2.0 to 2.6 quads in 2020 and 8.2 to 10.7 quads in 2030. However, even if such increases in vehicle fuel economy do occur, growing amounts of personal travel by automobile will likely cause overall emissions from cars and light trucks to increase in the coming decades. EIA (2009) projects that light-duty vehicles will use 16.53 quads of energy in 2030 compared to 16.42 quads today. Thus, additional opportunities for reducing GHG emissions must be considered. This includes direct efforts to reduce travel demand (discussed in the previous section) and expanded use of alternative fuels (discussed in the following section), as well as strategies for increasing efficiency in other areas of transport sector (discussed in Box 3.3).

BOX 3.3
Other Potential Opportunities for Increasing Transport-Sector Efficiency

Public Transit. Public transit is often cited for its potential to reduce automobile dependence and resulting energy use. The nation's transit systems, consisting of buses, light railways, ferries, rapid rail, and commuter lines, currently account for only about 3 percent of total U.S. passenger miles—but a more meaningful share of personal travel in a number of large metropolitan areas. The true energy efficiency of public transit depends upon the density of its use. In most places, transit is most heavily used for commuting during rush hours, and the systems run with low occupancy for much of the day. The net result is low levels of energy efficiency in many systems during these periods of low occupancy. Therefore, in trying to induce a mode shift to mass transit as a GHG emissions-reduction strategy, a challenge is in generating ridership throughout the day and drawing traffic primarily from single-occupant vehicles. To do so may require changes in the design and operations of transit systems, for instance, through the use of innovative, more flexible public transit concepts such as taxi buses.[1] At a more fundamental level, such changes need to coincide with policies aimed at influencing the location of housing and businesses in ways that make transit a more effective and appealing option for all kinds of local travel, not just work commuting (see Box 3.1 for further discussion of these transportation and land-use planning connections).

Aviation. Domestic air transportation currently accounts for about 12 percent of U.S. passenger miles and ~3 percent of total U.S. GHG emissions. Technological advances coupled with a highly competitive airline industry have prompted air carriers to steadily improve (by ~1-2 percent per year) the energy efficiency of their fleets and their operations during the past 40 years.[2] New aircraft

---

[1] Taxi bus refers to a mode of transport that falls between a private taxi and a conventional bus, often with a fixed or semifixed route, but without a fixed time schedule and with the capacity to stop anywhere to pick up or drop off passengers. They are currently being experimented with in some European cities.

[2] Schafer et al. (2009) found that, between 1959 and 1995, average new aircraft energy intensity declined by nearly two-thirds. Of that decline, 57 percent was attributed to improvements in energy efficiency, 22 percent resulted from increases in aerodynamic efficiency, 17 percent was due to more efficient use of aircraft capacity through higher load factors, and 4 percent resulted from other changes, such as increased aircraft size.

Reducing the Carbon Intensity of Energy

Opportunities for reducing the carbon intensity of energy include switching from higher- to lower-carbon-content fossil fuels, advancing coal technologies such as gasification and combined-cycle plants, along with carbon capture and storage (CCS) technologies, and advancing renewable energy sources (e.g., wind, hydro, geothermal, and solar power) and other no- or low-carbon-content energy sources such as hydrogen, nuclear, and biomass. Each of these opportunities is briefly discussed below.

entering the fleet today, such as the Boeing 787, use about 20 percent less fuel per seat-mile than the aircraft they are replacing. Airline fleet turnover should thus, over time, lead to continued gains in aircraft energy efficiency. At the same time, however, domestic airline passenger traffic is forecast to grow by between 2 and 3 percent per year over the next two decades, which is likely to offset the technological efficiency gains. Additional efficiency improvements can be found in navigation and control technology and changes in air traffic management practices. For instance, more direct routing and reduced taxiing and idling at airports represent potentially important areas of opportunity for future energy savings.[3]

Freight Transport. Rail transport currently accounts for about 40 percent of freight transport ton-miles, and is 5-15 times more energy efficient than trucks per ton-mile. Trucks account for about half of freight ton-miles, but a much higher share of certain (high-value, lighter-weight, nonbulk) shipments. Diverting more truck traffic onto rail presents a modest but real opportunity for energy savings. The potential is greatest for shipments going more than 500 miles, and this is already being exploited by some national truckload motor carriers, by placing trailers on rail cars for the line-haul segment of their trips. About 5 to 10 percent of truck traffic may be candidates for additional movement by rail. The DOE forecasts energy-efficiency improvements of 0.6 percent per year for heavy trucks for the next two decades (EIA, 2009). The 2007 EISA requires that the U.S. Department of Transportation (USDOT) establish fuel economy standards for medium- and heavy-duty trucks. An NRC committee is currently studying for USDOT the technological potential for energy-efficiency improvements in trucks in support of standards development. The scope, stringency, and structure of the EISA standards are not known at this time, but they may prompt energy-efficiency gains beyond those forecast by EIA (2009).

---

3 An additional factor to consider regarding aircraft is that the contrails emitted by planes at cruise level can have short-lived but potentially significant radiative forcing impacts (either warming or cooling, depending on altitude and related influences on cloud formation).

Natural Gas

Natural gas is the cleanest of the fossil fuels, with the lowest GHG emissions per unit of energy, emitting about half of the CO_2 of coal when burned for electricity generation. Shifting a greater fraction of coal- and petroleum-based energy use to natural gas is thus one potential means of reducing (although not eliminating) our nation's rate of CO_2 emissions growth. Currently, ~86 percent of the natural gas consumed in the United States is produced domestically, with much of the remainder coming from Canada. U.S. natural gas reserves have increased significantly over the past decade, largely because new technology has increased the accessibility of unconventional resources, especially gas shales. It is difficult to predict the long-term prospects to meet an increasing demand for natural gas, but recent estimates from EIA (2009) show that natural gas production from unconventional resources is growing. According to the AEF report, new natural gas combined-cycle power plants with CCS can compete economically with new coal plants with CCS, but the price of natural gas greatly affects this competitiveness. At the low price of $6/GJ, electricity costs are estimated at 7-10 cents/kWh. At $16/GJ, electricity costs are 14-21 cents/kWh. In comparison, electricity from coal with CCS is estimated at 9-15 cents/kWh (NRC, 2009a).

Renewables

Renewable-energy technologies that do not emit GHGs are an important and viable part of a near-term strategy for limiting climate change, and they could potentially play a dominant role in global energy supply over longer time scales, especially given the finite lifetimes of the fossil fuel-based options mentioned above. There is a wide variety of options for expanding the use of renewable energy resources (e.g., solar, wind, geothermal, biomass, and wave/tidal power). We do not review the current technical status of individual technologies here, as this has been done in detail in other recent reports, including *America's Energy Future: Real Prospects for Energy Efficiency in the United States* (NRC, 2009a) and *ACC: Advancing the Science of Climate Change* (NRC, 2010a). But we note some of the key challenges that the AEF report highlighted regarding the widespread expansion of intermittent renewable sources such as wind and solar.

For instance, AEF pointed to the technical challenges of increasing wind turbine capacity factors, lowering the cost of concentrating solar power through advances in high-temperature and optical materials, and developing increasingly thinner photovoltaic films at lower cost. Increasing the use of wind and solar for electricity will also require overcoming the challenge of integrating them into the grid. There is a need not only for greater transmission capacity but also for the increased installation of fast-responding generation to provide electricity when renewables are not available. Expanding the transmission system, improving its flexibility through advanced control technologies, and co-siting with other renewable or conventional generation facilities can help this integration.

Overall, the AEF report judged as feasible the goal of producing 20 percent of U.S. electric power from renewable sources by 2020[4] but not without substantial increases

---

[4] This included existing hydropower sources but not new additional ones.

in manufacturing capacity, employment, and capital investment. As a practical matter, local opposition to the siting of renewable electricity-generating facilities (such as wind farms) and associated transmission lines can also present barriers (as is true, of course, for other energy technology options as well). In order to facilitate investment in the face of high costs and risks, the AEF report observed that early and lasting commitments from policy makers are essential, including efforts to support research and development (R&D) and to avoid the "on/off" nature of federal tax credit programs (discussed further in Chapter 4).

Nuclear Power

Nuclear power is one of the key options for meeting large-scale electricity demand without producing GHGs. But the benefits of nuclear power must be weighed against a number of potential challenges. Strong public opposition to nuclear power, first evidenced in the 1970s, is rooted in a variety of concerns that any expansion of nuclear generating technology will confront (Rosa and Clark, 1999; Rosa and Dunlap, 1994; Whitfield et al., 2009). First is the challenge presented by the disposal of radioactive waste (particularly used fuel). The absence of a policy solution for the disposal of long-lived nuclear wastes, while not technically an impediment to the expansion of nuclear power, is still a public concern. New reactor construction has been banned in 13 U.S. states as a result, although several of these states are reconsidering their bans.

Safety and security concerns stem from the potential for radioactive releases from the reactor core or spent fuel pool following an accident or terrorist attack. Nuclear reactors include extensive safeguards against such releases, and the probability of one happening appears to be very low. Nevertheless, the possibility cannot be ruled out, and such concerns are important factors in public acceptance of nuclear power. Proliferation of nuclear weapons is a related concern, but after 40 years of debate, there is no consensus as to whether U.S. nuclear power in any way contributes to potential weapons proliferation. A critical question is whether there are multilateral approaches that can successfully decouple nuclear power from nuclear weapons (Socolow and Glaser, 2009). Other potential barriers to the deployment of new nuclear plants include the high capital costs of building new plants as well as the time-consuming and costly permitting, certification, and licensing processes.

Nuclear plants now in place in the United States were built with technology developed in the 1960s and 1970s. In the intervening decades, ways to make better use of existing plants have been developed, along with new technologies that improve safety and security, decrease costs, and reduce the amount of generated waste—especially high-level waste. These technological innovations include improvements or modification of existing plants, alternative new plant designs (e.g., thermal neutron reactor and fast neutron reactor designs), and the use of alternative (closed) nuclear fuel cycles. These new technologies under development may allay some of the concerns and barriers described above, but it will be necessary to determine the functionality, safety, and economics of those technologies through demonstration and testing.

Considering only technical potential (i.e., not accounting for the practical barriers discussed above), AEF estimates that a 12-20 percent increase in U.S. nuclear capacity is possible by 2020. After 2020, the potential magnitude of nuclear power's contribution to the U.S. energy supply is uncertain and will depend on the performance of plants built during the next decade.

Biopower and Biofuels

 Biopower for electricity and biofuels for transportation can be produced from many sources, including wood and plant waste, municipal solid waste and landfill gas, agricultural waste, and energy crops. The AEF report concludes that, between now and 2020, there are no technological constraints to expanding biofuels production using existing technologies, but at high levels of deployment other types of barriers arise. These include the challenge of producing the fuel near enough to generating facilities to make hauling feasible and producing biomass feedstock in a sustainable manner that avoids excessive burdens on ecosystems.

The AEF committee considered corn grain ethanol to be a transition fuel to cellulosic biofuels or other biomass-based liquid hydrocarbon fuel. Biochemical conversion of grains to ethanol has already been deployed commercially and was important for stimulating public awareness and initiating the industrial infrastructure. There is active debate, however, about the land-use and GHG implications of greatly expanding biomass-based energy sources. For instance, Melillo et al. (2009) predict that an expanded biofuels program will significantly increase direct carbon losses from soil, as well as indirect losses resulting from expanded conversion of forests and grasslands, and from the additional nitrous oxide emissions from increased fertilizer use. Searchinger et al. (2008) likewise argue that the land-use changes that might occur internationally as a result of diverting significant U.S. corn crops to ethanol production would lead to a large net increase in GHG emissions. Others, however (e.g., Wang and Haq, 2009), assert that such claims are based on flawed assumptions. In response to such concerns and uncertainties, a growing number of groups are developing standards for assessing and certifying the sustainability of processes for obtaining biopower feedstocks (e.g., the Roundtable on Sustainable Biofuels; see *http://cgse.epfl.ch*).

Cellulosic ethanol and other advanced cellulosic biofuels have much greater potential to reduce U.S. oil use and CO_2 emissions, while having smaller impacts on the food supply. The AEF report suggests that cellulosic biomass—from dedicated energy crops, agricultural and forestry residues, and municipal solid wastes—could potentially be produced on a sustainable basis using today's technology and agricultural practices. The time frame for technology development and deployment is uncertain, however.

Carbon Capture and Storage

CCS could be used to remove CO_2 from the exhaust gases of power plants fueled by coal, natural gas, or other carbonaceous material (including biomass), as well as from other large industrial processes that emit CO_2 (such as natural gas processing or the production of hydrogen, ammonia, and other chemicals).[5] CCS technologies have been demonstrated at the commercial scale in several large industrial processes, but no large power plant today captures and stores its CO_2. Although current technologies can capture roughly 90 percent of CO_2, they are quite costly in power plant applications. CO_2 storage (or sequestration) could be implemented most effectively in several types of geological formations, including depleted oil and gas reservoirs and beneath other layers of impermeable rock. Specific sites would have to be selected, engineered, and operated with careful attention to safety. In particular, the deep subsurface rock formations that trap the CO_2 must allow the injection of large quantities and ensure its containment over timescales of centuries.

More reliable cost and performance data are needed both for capture and storage, and these data can be obtained only by construction and operation of full-scale demonstration facilities. Such demonstrations could provide vendors, investors, and other private-industry interests with the confidence that power plants incorporating advanced technologies, and the associated storage facilities, could be built and operated in accordance with commercial criteria. Similarly, to sort out storage options and gain experience with their costs, risks, environmental impacts, legal liabilities, and regulatory and management issues (e.g., Pollak and Wilson, 2009; Wilson et al., 2007), it will be necessary to operate a number of large-scale capture and storage projects that encompass a range of different fuels (coal, natural gas, and biomass), application types

---

[5] Because the CO_2 is captured before it is emitted to the atmosphere, it is classified here as an option for "reducing carbon intensity" rather than as "post-emission carbon management."

(e.g., pre- and postcombustion), and geological formations. And like any technology, favorable public perception and social acceptance may prove to be crucial for widespread deployment of CCS. A small number of studies are beginning to explore such issues (e.g., Bielicki and Stephens, 2008).

The investments needed to create this portfolio of CCS demonstrations will certainly be significant—approximately $1 billion per project for large coal plants—but there is no benefit in waiting to make such investments. The AEF committee judges that the period between now and 2020 could be sufficient for acquiring the needed information on CCS viability, provided that the deployment of CCS demonstration projects proceeds as rapidly as possible; if these investments are made now, 10 gigawatts (GW) of CCS projects could be in place by 2020.

Post-Emission Carbon Management

Preventing or limiting GHG emissions from known sources is the classic abatement approach that dominates most current policy deliberations. However, fossil fuels will remain abundant and relatively inexpensive for many years to come; there is little evidence thus far that most nations of the world are willing to take maximum advantage of the GHG emissions-reduction opportunities discussed in the previous sections. There is strong motivation, therefore, to consider complementing traditional emissions-reduction efforts with strategies for managing carbon after it is already released into the atmosphere—to extract CO_2 from the air and keep it sequestered in a stable reservoir for at least decades to centuries. This includes opportunities such as enhancing natural biological sequestration processes (e.g., in forests, soils, and the ocean surface) and developing chemical or mechanical sequestration processes (e.g., capturing CO_2 with chemicals or materials). Some of these strategies are already well characterized and widely used, while others are in an early stage of conceptualization. Depending on the nature and scale of such efforts, they are sometimes referred to as forms of "geoengineering" (see Box. 3.4).

Sequestration in Forests and Agricultural Systems

Reducing emissions from deforestation and forest degradation (referred to as REDD in policy circles) can play an important role in global efforts to limit the magnitude of climate change, because this is the source of roughly 17 percent of current global GHG emissions (IPCC, 2007a). REDD is viewed as a major target for international emissions offset opportunities, since the costs (per ton of CO_2 emissions) is much less than emissions reductions taken in the energy sector of industrialized countries, and it can

---

BOX 3.4
The Panel's Approach to Geoengineering

There is growing interest in strategies for deliberate, large-scale manipulations of the Earth's environment in order to offset the harmful consequences of GHG-induced climate change. These proposed manipulations are generically referred to as "geoengineering." Although few are promoting geoengineering as an alternative to traditional emissions-reduction strategies, the concept has recently been gaining more serious attention as a possible "backstop" measure. This attention is driven by lack of progress in making a large-scale transition to low-GHG technologies and by growing evidence that the world may be on a dangerous trajectory with respect to climate change, regardless of whether there is substantial future progress in limiting GHG emissions.

One broad category of proposed geoengineering schemes encompasses methods for "solar radiation management," such as injecting sulfate particulates into the stratosphere, enhancing global cloud cover, and other methods of affecting the reflectivity of Earth's surface or atmosphere. As discussed in *ACC: Advancing the Science of Climate Change* (NRC, 2010a), solar radiation management schemes are fraught with scientific uncertainties about the efficacy of intended impacts and the possibility of unintended impacts, as well as questions about ethical implications, public acceptance, and international governance.

This report considers only schemes for removing CO_2 directly from the atmosphere (that is, "post-emission carbon management"), which in general do not raise the distinctively difficult governance, ethical, or public acceptance issues raised by other forms of geoengineering.[1] In addition, post-emission carbon management schemes can ultimately help to address another major environmental concern associated with growing CO_2 emissions: increasing acidification of the world's oceans (also discussed in *ACC: Advancing the Science of Climate Change* [NRC, 2010a]).

---

[1] One possible exception is ocean fertilization, which is discussed later.

offer a variety of ancillary ecological benefits for developing countries (discussed in Chapter 6).

In terms of domestic action, the United States can augment its emissions-abatement efforts with a variety of practices to enhance carbon sequestration in its own forests and croplands. This includes planting new forests (afforestation), protecting existing forests against loss and degradation, reducing cropland tillage, and enhancing conversion to grasslands. McCarl and Schneider (2000) and McCarl and Reilly (2007) provide details on specific sequestration mechanisms and discuss how these sorts of efforts affect agricultural emissions of GHGs (CO_2, CH_4, N_2O), as well as the use of land for growing biofuel feedstock.

Studies by the Environmental Protection Agency (EPA) (Murray et al., 2005) conclude that the national emissions reduction capacity of forestry and agricultural practices is ~630 teragrams (Tg) CO_2/yr in the first decade, declining to about 85 Tg CO_2/yr by 2055 due to saturation of carbon sequestration (discussed below) and carbon losses after timber harvesting. Also, despite the declining annual sequestration rates, cumulative agricultural GHG emissions steadily increase over time (assuming that food production remains a priority in land-management practices). For a scenario with a constant price of $15 per ton CO_2-eq, the cumulative amount reaches ~26,000 Tg CO_2-eq by 2055.

Different types of land-use practices have different mitigation potentials, as illustrated by Figure 3.2, with forestry practices generally demonstrating much larger sequestration capacity than agricultural soil management. Optimal carbon-sequestration strategies are largely a function of time and GHG price levels. For instance, Murray et al. (2005) found that, at relatively low GHG prices ($5 per ton CO_2-eq or less) and in early years, carbon sequestration in agricultural soils and in forest management would be optimal as the dominant mitigation strategies; at middle to higher prices ($15 per ton CO_2-eq or higher) in the early to middle years, afforestation becomes the leading strat-

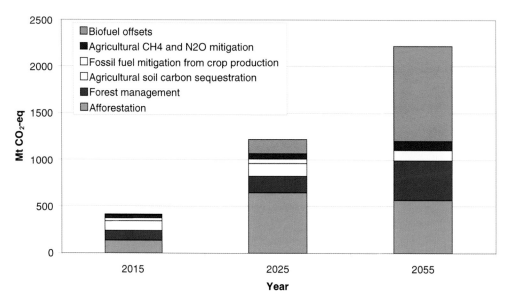

FIGURE 3.2 Mitigation potential in the U.S. agriculture and forestry sectors, assuming a price of $20 per ton CO_2-eq in 2010, increasing by $1.30/yr. (The negative value in 2055 indicates less sequestration relative to the baseline value.) Note that both the absolute and relative magnitudes of different sequestration options vary over time. SOURCE: EPA.

egy; and at the highest prices ($30 and $50 per ton CO_2-eq) and in years beyond 2050, biofuels (primarily for electricity) dominate the portfolio.

It is important to recognize that the physical volumes of sequestration implied by some estimates potentially neglect the costs of adoption and the competition for land from alternative uses. McCarl and Schneider (2001) found that, even at very high prices (up to $500 per ton CO_2-eq), the *economic potential* to sequester carbon in agricultural soils is less than two-thirds of the pure *technical potential*. Furthermore, at such high prices, soil sequestration has to compete with other strategies such as bioenergy and afforestation, which further reduce the *competitive economic potential* for soil carbon sequestration to less than one-third of the technical potential.

One constraint to keep in mind is that carbon sequestration practices are effective only until the biological capacity of the ecosystem is reached and the system becomes saturated for a land use, that is, when the rate of carbon additions to the ecosystem reaches equilibrium with the rate of decomposition and re-release of carbon to the atmosphere. In the case of soil tillage, this typically happens within 10 to 15 years; grassland conversions can continue sequestration for 30 to 50 years; and in forestry it can exceed 80 years. Some new approaches for alleviating saturation efforts are being explored (Box 3.5), but in general, these sorts of biological sequestration strategies are best viewed as near-term bridging strategies for helping to manage atmospheric GHG concentrations during the time required to ensure widescale implementation of more long-term (i.e., energy-sector-based) solutions. There are a number of other practical constraints that make these forms of post-emission carbon management more challenging than traditional emissions-reduction efforts. For instance:

- *Transaction costs and land unit size.* Many of the strategies to generate GHG emissions reductions typically involve small volumes for a given landowner per unit. For example, carbon sequestration on agricultural soils generates about 0.8 tons CO_2 per acre of land and the average farm size is about 450 acres. This means a 100,000-ton offset would require a group of almost 280 farmers. This raises the issue of relatively high transaction costs in assembling the group and measuring, monitoring, and verifying practices.
- *Leakage.* A number of the proposed agricultural and forestry options, such as biofuels production and afforestation, divert land from conventional production of agricultural commodities. If this then diverts commodities from the marketplace, it can lead to changes in land use such as deforestation or conversion of grasslands into agricultural production elsewhere in the world, with accompanying emissions increases (Fargione et al., 2008;

BOX 3.5

Emerging Strategies for Enhancing Biological Carbon Sequestration

Post et al. (2009) identify a number of approaches that may alleviate the ecosystem satura-tion effects that currently limit carbon sequestration potential, for instance:

Biotechnology. The postgenomics era provides an opportunity to identify genes, enzymes, and other factors that underlie rate-limiting steps in carbon acquisition, transport, and fate and potentially open up new approaches to enhance terrestrial carbon sequestration.

Deep-soil sequestration. Carbon decay in undisturbed soil at depths from 0.2 to 3 m is mini-mal. To use this reservoir as an efficient sequestration pool, mechanisms need to be developed and adopted for moving carbon into these lower soil depths. Amending soils with lime, urea, and phosphate fertilizers offer one such approach, as do planting deep-rooted perennials.

Biochar. Biochar burial involves creating a charcoal-like substance by pyrolyzing[1] harvested biomass in a process that renders it inert. The biochar can then be used as a soil amendment to improve soil fertility, increase crop productivity, and provide additional carbon sequestration ben-efits (Lehmann et al., 2006; McHenry, 2009). This is because the carbon contained in the biochar is unavailable for oxidation to CO_2 and subsequent release to the atmosphere. Conversion of biomass carbon to biochar leads to initial sequestration of about 15 to 35 percent of the carbon released by the initial feedstock biomass. A preliminary appraisal finds that at present this approach is costly, and significant uncertainty remains about long-term sequestration effectiveness, but it does have the potential to yield a negative carbon balance (McCarl et al., 2009).

---

[1] Pyrolysis is a form of incineration that chemically decomposes organic materials by heat in the absence of oxygen.

Searchinger et al., 2008). These sorts of emission leakage issues are dis-cussed further in Chapter 4.

- *Additionality.* Many of the agricultural practices that could be stimulated by a GHG price or tax are already in use, and thus it may be questioned whether they are "additional." A widely held stance is that credits should be granted for GHG offsets that are additional to what would have occurred under business as usual. Yet farm groups want to reward good actors who started a practice before the pricing program began. This raises major policy design challenges. For example, how should policy deal with payments to preexisting practices, or how can new actions be distinguished from those that would have oc-curred anyway?

- *Permanence and uncertainty.* The agriculture and forestry sectors are character-ized by pervasive uncertainty in terms of year-to-year fluctuations in commod-ity yields. And, as noted above, sequestered carbon may not be permanent.

The possibility of sequestration reversal raises concerns about how offsets might be treated in the marketplace.

- *Property rights, commitments, and leasing.* Farm and forestry groups have reservations about making permanent commitments to carbon sequestration. Many such groups favor leasing rather than permanent sales, because they are worried about factors such as future carbon prices; requirements that land be managed in particular ways; potential increases in cost, particularly for weed and insect control; and critical reliance on the efficacy of chemical weed control compounds, in the face of possible development of resistance to control methods. Furthermore, rules to prevent converting grasslands into croplands would infringe on private property rights (Marland et al., 2001), which could lead to major legal obstacles to program implementation and management, as well as associated transactions costs.

- *Measurement and monitoring.* Implementing markets for agricultural practices requires rigorous measurement and monitoring protocols. Some argue that these elements will be quite costly, but others argue that standard soil-sampling methodology and process-based modeling offer low-cost approaches (Mooney et al., 2004; Paustian et al., 2009), especially because a 5- to 10-year period of time between sampling will likely be required to detect changes (Conant and Paustian, 2002; Smith, 2004).

- *Payment by practice versus by outcome.* Many soil and forest programs are targeted to reward practices, not environmental outcomes such as the amount of carbon sequestered. Antle et al. (2001) indicated this is inefficient on a per-ton cost basis, because any particular practice can lead to widely varying outcomes in terms of carbon sequestered. Wu and Boggess (1999), however, argued that per-acre or per-practice policies are more efficient than policies based on tons of carbon sequestered or erosion avoided.

Oceanic Sequestration Strategies

One of the methods that have been proposed to enhance natural biological carbon uptake is iron fertilization of the oceans. This involves the intentional introduction of iron to upper ocean waters to stimulate a phytoplankton bloom, with the goal of enhancing biological productivity (the growth of plant biomass) and enhanced removal of CO_2 from the atmosphere. A number of research groups have conducted preliminary studies of this strategy in theoretical, laboratory, and ocean field tests. Ocean trials have demonstrated that phytoplankton blooms can be stimulated by iron addition (Boyd et al., 2007), but much controversy remains over the effectiveness of this method for atmospheric CO_2 sequestration and over its effects on ocean ecology and biology (Buesseler et al., 2008). In addition, any sort of large-scale ocean manipulation scheme would likely need to be carried out as an international cooperative effort and thus faces major political and institutional hurdles. The Convention on the Prevention of Marine Pollution by Dumping of Wastes and Other Matter 1972 (London Convention, 1972) has already been involved in developing governance for experimental studies in this arena.

Geochemical Sequestration Strategies

A number of strategies have been proposed for using geochemical reactions that enhance transformation of CO_2 gas into dissolved or solid-phase carbon. Stephens and Keith (2008) reviewed the technical status of these geochemical approaches, grouped into three broad categories:

1. *Mineral carbonization.* Various mechanical methods have been proposed to accelerate the naturally slow reactions wherein CO_2 in the atmosphere is converted to carbonates and returned to the lithosphere by weathering of rocks.
2. *Altering ocean alkalinity.* Deliberately increasing ocean alkalinity through various geochemical means may increase the ocean's capacity to store dissolved inorganic carbon and also help address the problem of increasing ocean acidity.
3. *In situ geochemical processes.* One can facilitate geochemical reactions in locations where minerals exist naturally, including enhanced carbonate formation in calcium and magnesium silicate-rich aquifers, and carbonate dissolution in submarine carbonate deposits.

Stephens and Keith (2008) suggested that more research is needed to effectively compare the economic viability of the different approaches, but at present, among these geochemical approaches, only alkalinity addition seems to provide a significant improvement beyond conventional CCS in broadening the economic scope of carbon storage options.

Direct Air Capture of CO_2

For large point sources such as power plants, on-site CO_2 capture from an effluent stream is considered technically feasible and potentially cost-effective. But for distributed sources such as vehicles and small industrial sources (which account for nearly half of all GHG emissions globally), this type of capture method is not technically or economically feasible. Thus, one strategy for dealing with these dispersed emissions sources involves finding the means to extract CO_2 directly from the ambient air. This direct-capture strategy is appealing for numerous reasons: it can be colocated with suitable geological storage sites; it eliminates the need to ship captured CO_2 from its source to a disposal site; it could be deployed as soon as it is developed (i.e., one would not have to wait for the phase-out of existing energy infrastructure to begin implementation); and it could likely be carried out at the national level, without the need for new international agreements or governance institutions.

One class of strategies for direct air capture that has emerged thus far involves physical or chemical absorption from airflow passing over some recyclable sorbent such as sodium hydroxide. A few research groups are developing and evaluating prototypes of such systems (Keith et al., 2006; Lackner et al., 1999). Major challenges remain in making such systems viable in terms of cost and energy requirements and improving overall capture energy efficiency. And of course, the challenges of long-term storage of the captured CO_2 are the same as those discussed earlier for CCS from industrial sources. If the technology were to someday become technically and economically feasible, however, the amount that could be captured would face no physical limit (other than global storage capacity) and, thus, could fundamentally alter the picture for efforts to reduce atmospheric GHG concentrations.

Reducing Emissions of Non-CO$_2$ Greenhouse Gases

Roughly 15 percent of U.S. GHG emissions (based on CO_2 equivalents) come from non-CO_2 gases, including methane (CH_4), nitrous oxide (N_2O), and fluorinated industrial gases such as hydrofluorocarbons (HFCs), perfluorocarbons (PFCs), and sulfur hexafluoride (SF_6) (EPA, 2009). Pursuing non-CO_2 GHG emissions-reduction opportunities can be an attractive option because these gases are, per molecule, generally much stronger climate forcing agents than CO_2; studies have shown that including non-CO_2 emissions-reduction options allows involvement of a far wider and more diverse set of economic sectors and opportunities, leading to a substantial reduction in the overall economic cost of limiting GHGs (e.g., Clarke et al., 2009; de la Chesnaye et al., 2007).

There are technically feasible strategies for reducing some non-CO_2 GHG emissions at negative or modest incremental costs. Many of these strategies are discussed briefly below, and more detailed discussion can be found in the literature (EPA, 2006, and see Table 3.1). Note that some strategies can yield multiple environmental benefits; for example, later in this chapter we discuss the example of controlling chemical species that affect both climate change and air quality (e.g., black carbon, tropospheric ozone,

TABLE 3.1 Summary of Emission Reduction Options for Non-CO$_2$ GHGs

Gas	Source	Key Opportunities	Estimated Emissions Reduction Potential (Mt CO$_2$-eq for 2030)
CH$_4$	Landfills	Methane recovery and combustion (i.e., power generation, industrial uses, flaring)	92
	Coal mines	Methane recovery and combustion, flaring, ventilation air use	39.2
	Oil/gas systems	Use of low-bleed equipment, better management practices	47.5
	Livestock waste	Methane collection from anaerobic digestors and combustion (power, flaring)	9
	Ruminant livestock	Improved production efficiency through better nutrition and management	12
	Rice production	Water management, organic supplements	4
N$_2$O	Industrial sources	Adipic acid (catalysts, thermal destruction), nitric acid (nonselective catalytic reduction)	25.9
	Mobile sources	Noncombustion vehicle alternatives (i.e., electric cars, fuel cell vehicles), reduced existing vehicle use (public transportation, fuel efficiency)	??

and methane). We note also the example that measures taken under the Montreal Protocol to control emissions of compounds that deplete stratospheric ozone have been estimated to create a climate change benefit (at zero incremental cost) that is five to six times what would have occurred if the Kyoto Protocol had been fully implemented (Velders et al., 2007).

TABLE 3.1 Continued

Gas	Source	Key Opportunities	Estimated Emissions Reduction Potential (Mt CO_2-eq for 2030)
	Soil management	Precision agriculture, cropping system models, controlled-release fertilizers, soil conservation practices	36.2
SF_6	Aluminum production	Reduced anode effects	0.8
	Magnesium production	Improved process management, SF_6 substitutes	0.9
	Electric power	Improved gas handling, recycling, new equipment	3.6
	Semiconductors	Improved process management, thermal destruction, alternative chemicals	1.5
HFC-23		Improved process management, thermal destruction	2.9
Ozone-depleting substance substitutes		Improved gas management, alternative chemicals, banning of nonessential uses	84.9

SOURCES: Nonagricultural technical potential estimates from EPA (2006) and EPA's legislative analyses. Estimates assume a price of $60/ton CO_2-eq. Agricultural estimates are from the results of Baker et al. (2009) at a $50 CO_2-eq price. Description of emissions-reduction opportunities from EPA (see *http://gcep. stanford.edu/pdfs/3KC3dzpRALy3cHpkGrwJCA/Paul_Gunning_Non-CO₂.pdf*).

Methane Emissions from Energy and Landfill Sources

Methane (CH_4) is emitted from leaks or venting from oil and gas systems, landfills, and coal mining. Reducing these emissions is cost-effective in many cases, due to the market value of the recovered gas.[6] Cost-effective CH_4 emission-reduction technologies and practices (e.g., leak detection and reduction activities) already exist, but there is

---

[6] If one uses the captured methane as a fuel, and this displaces the use of more carbon-intensive fuels, it is a net gain in terms of GHG emissions.

opportunity for their broader deployment. Reductions in *landfill gases* will result from deployment of existing approaches, coupled with waste reduction through recycling, and technological advances in solid-waste management technologies. For *coal mine emissions*, deployment of existing emissions-control techniques, advances in coal mine ventilation, and new coalbed methane drilling techniques would all help to further reduce emissions. Currently, U.S. industries and state and local governments collaborate with the EPA in a variety of voluntary programs to promote and overcome informational, technical, and institutional barriers to reducing CH_4 emissions. EPA expects that these programs will maintain emissions below 1990 levels in the future. Through the EPA Methane to Markets Partnership, the United States is also working toward reducing international CH_4 emissions.

Nitrous Oxide Emissions from Combustion and Industrial Sources

N_2O emissions from combustion and industrial acid production accounted for nearly 7 percent of U.S. non-CO_2 GHG emissions. Combustion of fossil fuels by mobile (e.g., trucks, cars, buses, trains, and ships) and stationary (steam boilers and other systems used for power and heat production) sources is the largest nonagricultural contributor to N_2O emissions. For the combustion sources, N_2O emissions appear to vary greatly with different technologies and operating conditions. Current research is aimed at identifying the most promising approaches and technologies for reducing N_2O emissions from these sources, but no technologies are suitable for deployment at this time. The largest industrial source of N_2O emissions is from nitric acid production. Virtually all of the nitric acid produced in the United States is manufactured by the catalytic oxidation of ammonia; therefore, development of advanced catalysts could further limit N_2O emissions from this source.

Methane and Nitrous Oxide Emissions from Agriculture

The largest overall source of non-CO_2 GHG emissions is from agriculture, in particular CH_4 from enteric fermentation in ruminant livestock and N_2O and CH_4 from manure and fertilizer. A number of methods and technologies are available today to help reduce some of these emissions. For instance, advanced imagery, precision agriculture, and sensing and control technologies are available to help farmers minimize overfertilization practices that lead to emissions and to apply fertilizers under conditions that decrease transformation of fertilizer nitrogen into N_2O. New chemical fertilizers that minimize gaseous losses and inhibit nitrogen transformation to N_2O are also available. CH_4 emissions from manures can be greatly reduced by improving livestock waste management systems through use of anaerobic treatment and gas recovery systems (commonly called anaerobic digesters). Methane from enteric fermentation can also be reduced somewhat through better feed and forage management, breed improvements, diet management, and strategic feed selection.

Fluorinated GHG Emissions

Although emissions of fluorinated GHGs are relatively small, contributing only about 2 percent to total CO_2-eq emissions, their 100-year global warming potentials (GWPs) are significant, ranging from 124 to 22,800 times that of CO_2. Emissions-reduction options address three categories of emissions:

1. *Unintended by-products.* There are two sources of unintended by-product emissions: PFC emissions (primarily PFC-14) from aluminum production and hydrofluorocarbon-23 (HFC-23) emissions from hydrochlorofluorocarbon-22 (HCFC-22) production. From 1990 to 2007, voluntary industry programs for improving process and control measures reduced PFC emissions from 19 to 4 million metric tons (MMT) CO_2-eq. Achieving further significant reductions, however, would require major new advances in these processes. From 1990 to 2007, HFC-23 emissions were reduced from 36.0 to 17 MMT CO_2-eq, through a combination of process optimization and capture and destruction of the compound. Virtually all of the remaining emissions could be eliminated, at costs estimated to be as low as $0.20 per metric ton.

2. *Intentionally produced compounds.* Some GHGs are intentionally produced for use in a wide range of consumer and commercial applications and are used in billions of pieces of equipment and products worldwide. Their emissions can occur years to decades after production, which makes downstream emissions control very difficult. Of this class, the most significant are HFCs, used as replacements for ozone-depleting substances controlled under the Montreal Protocol (primarily in refrigeration and air-conditioning systems). U.S. consumption of HFCs in 2005 was estimated at 170 MMT CO_2-eq and consumption is projected to grow as HCFC consumption is reduced and HFCs are used in their place. Emissions-reduction options range from better refrigerant management to minimize emissions, to substitution with alternative fluids and technologies with lower GWPs (although options for the latter have yet to be identified in many cases). The potential exists for even greater growth in HFC use in developing countries, primarily due to rapidly increasing demand for refrigeration and air-conditioning. Velders et al. (2009) argue that developing-country use in 2050 could exceed that in industrialized countries by a factor of 8. Other uses for intentionally produced compounds are PFCs, SF_6, NF_3, and

HFCs for semiconductor manufacture, and SF_6 for electrical transmission and distribution and for magnesium production and processing. Voluntary programs for all these applications have succeeded in reducing U.S. emissions from about 35 MMT CO_2-eq in 1990 to 20 MMT CO_2-eq in 2007. Future reduction potential is uncertain.

3. *Capturing and destroying compounds.* Even though the consumption of some of the most widely used ozone-depleting substances (chlorofluorocarbons [CFCs]) has been phased out, significant banks of the compounds still exist in refrigeration and air-conditioning equipment and in insulating plastic foams. Destruction costs using approved technologies range from $2.75 to $11 per kg of the CFC, not accounting for the additional cost of recovery, storage, and transportation (IPCC/TEAP, 2005). Due to the high GWP of these compounds, their capture and destruction can be cost-effective (on a per ton CO_2-eq basis); for example, if it cost $10 to capture and $10 to destroy CFC-12, then the cost is approximately equivalent to $2 per ton CO_2-eq. The size of this emissions-reduction opportunity is rapidly diminishing with time as the remaining CFCs continue to leak from systems worldwide. Controlling these leaks would help mitigate ozone depletion and help limit the magnitude of future climate change.

Short-Lived Radiative Forcing Agents

Most discussion on limiting the magnitude of climate change focuses on "well-mixed" GHGs that persist in the atmosphere for periods ranging from years to centuries (even millenia, for the PFCs). Although less frequently mentioned in climate discourse, reducing atmospheric concentrations of short-lived atmospheric pollutants (namely, tropospheric ozone and black carbon particles) may offer a cost-effective near-term strategy for limiting the magnitude of climate change, while at the same time producing substantial benefits for air quality.

Tropospheric ozone (O_3) is itself a strong GHG, but it also plays a key role in atmospheric chemistry, affecting the lifetimes and hence the concentrations of several other important GHGs, including CH_4, HCFCs, and HFCs. O_3 is not emitted directly but is produced in the atmosphere via reactions among its precursors: nitrogen oxides (NO_x), carbon monoxide (CO), methane (CH_4), and nonmethane hydrocarbons. Thus, controlling O_3 requires controls on the emissions of these precursors. Some ozone precursor emissions (from sources such as vehicles, factories, power plants, consumer products, and paints) are currently controlled through provisions of the Clean Air Act. Since CH_4 is a precursor for O_3 formation on a broad regional level (as opposed to the context of concentrated urban air pollution), there are multiple reasons for pursuing strategies that reduce CH_4 emissions from industrial, energy, and agricultural systems. This would not only reduce the climate impacts of CH_4 itself but also help lower the climate impacts of O_3 (West et al., 2006).

Black carbon or "soot" not only causes strong direct warming in the atmosphere (on a localized scale) but also amplifies warming effects after deposition from the atmosphere because the resulting black coating on certain surfaces (such as arctic snow and ice) decreases the amount of incoming solar radiation these surfaces reflect back to space. Black carbon is emitted from the burning of fossil fuels, biofuels, and biomass. Diesel emissions account for 30 percent of black carbon globally and 50 percent in the United States. Technology for reducing soot emissions from diesel engines exists and is already mandated for new diesel vehicles in the United States. Reducing these emissions would have important domestic benefits for human health, but benefits at the international level are even more profound. For instance, replacing primitive biomass cookstoves that emit large amounts of soot with inexpensive, clean technologies could have enormous health benefits for the millions of people who suffer from this dangerous source of indoor air pollution (WHO, 2005).

Including these sorts of short-lived compounds in a larger GHG emissions-reduction effort does pose methodological challenges (for instance, it is difficult to apply the concept of GWPs and CO_2-equivalent emissions to such species). Nonetheless, it has been suggested that focusing on these short-lived species could be particularly advantageous as a near-term bridging strategy for easing climate change during the time required for major CO_2 emissions controls to come into play. It is especially attractive as an international strategy because low-income countries that view CO_2 emissions reduction as a threat to their economic growth often see the control of pollutants such as O_3 and soot as an immediate, obvious benefit. Also, because these short-lived pollutants are rapidly removed from the atmosphere, reducing emissions will have a near-immediate effect on lowering atmospheric concentrations.

THE CASE FOR URGENCY

Chapter 2 drew on the Energy Modeling Forum (EMF22) project to identify a representative domestic GHG emission budget range of 170 to 200 Gt CO_2-eq for the period 2012 through 2050, and earlier sections of this chapter identified a wide range of opportunities for reducing domestic GHG emissions. Here we assess whether the technical potential for domestic emissions reduction is sufficient to meet a domestic GHG budget in the suggested range (assuming, as discussed in Chapter 2, that international offsets are not used to meet the U.S. domestic GHG budget[7]). Based on this assessment, we conclude that meeting the representative U.S. GHG budgets may be feasible, but only if the nation acts with great urgency to deploy available technologies and to create new ones.

This conclusion is based on two analyses described below. First, we find that the energy efficiency and energy production technologies available for near-term commercial use (i.e., by 2020) could attain the deployment levels required for meeting the emissions budget scenarios only under the most favorable circumstances. Because the margin for error is so thin, meeting the budget using only these technologies seems unlikely. Second, we find that, without prompt action, the current rate of GHG emissions from the energy sector would use up the domestic emissions budget well before 2050. In short, meeting the emissions budget scenarios considered in Chapter 2 means that the United States needs to start decarbonizing its energy system as soon as possible but does not yet have in hand the suite of technologies needed to complete the task. We reiterate the point made in Chapter 2 that the U.S. emissions budget used in this analysis is based on "global least-cost" economic efficiency criteria and that credible political and ethical arguments can be made for a more aggressive U.S. effort than the one we discuss in this section. To meet these more ambitious targets would, of course, be even more difficult.

Feasibility of Decarbonizing the Energy System

To assess the feasibility of decarbonizing the energy system, we compare the possible requirement for future energy efficiency and energy supply technologies with the likely availability of those technologies. Two recent studies, EMF22 and AEF, provide the data to make this comparison directly. Figure 3.3 shows a set of scenarios developed in the EMF22 studies that illustrate the types of changes to the energy system that might be needed to reach an emissions budget of either 167 or 203 Gt CO_2-eq by 2050.[8] Below are the results of five different models, showing the energy technology mix projected for 2050 compared to the mix in the year 2000.

There are large uncertainties associated with these sorts of projections, but the varia-

---

[7] If the United States does rely heavily on the use of international offsets to meet an emissions budget, that would mean less stringent requirements for actually reducing domestic emissions; thus, the energy mix going forward would likely include a larger percentage of freely emitting fossil fuels than in the cases shown in Figure 3.3.

[8] As noted earlier, the EMF-22 analysis cases are 167 and 203 Gt CO_2-eq, which we rounded to 170 and 200 Gt CO_2-eq in Chapter 2. This difference does not significantly affect the conclusions of our analysis.

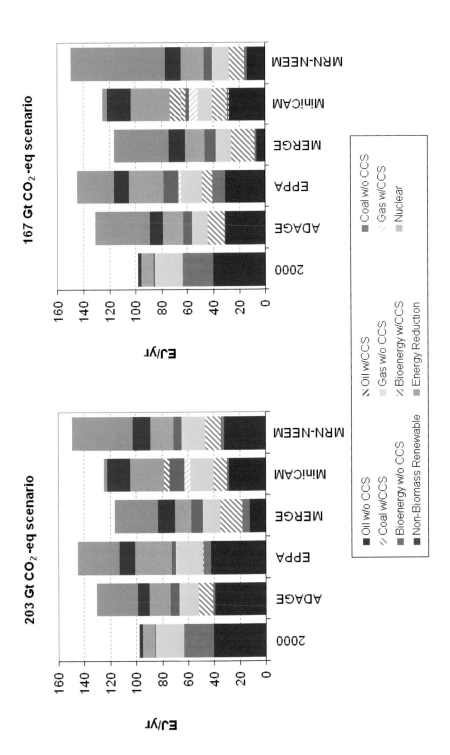

FIGURE 3.3 Model projections (from the EMF22 study) of the mix of energy technologies that may be used in 2050, under scenarios with emission budgets of 203 and 167 Gt CO_2-eq. For comparison, the first column in each graph shows the U.S. energy technology mix in 2000. A wide variety of future energy mix scenarios is possible, but all cases project a greater role for energy efficiency, renewable energy, fossil fuels with CCS, and nuclear power. SOURCES: Adapted from Fawcett et al. (2009); see also *http://emf.stanford.edu* for further details.

tion among them illustrates that the United States has many plausible options for configuring its future energy system in a way that helps meet GHG emissions-reduction goals. Note, however, that all cases involve a greater diversity of energy sources than exist today, with a smaller role for freely emitting fossil fuels and a greater role for energy efficiency, renewable energy, fossil fuels with CCS, and nuclear power. The virtual elimination by 2050 of coal without CCS—presently the mainstay of U.S. electric power production—in all the scenarios is perhaps the most dramatic evidence of the magnitude of the changes required.

The AEF study estimated the technical potential of the rate at which key technologies can be deployed over the next 25 years, based on the committee's judgments of when technologies will be available for commercial deployment and the likely maximum rate of deployment thereafter (see Box 3.6 for further explanation of "technical potential"). In Table 3.2, these AEF technical potential estimates are compared with estimates from EMF22 studies of the technological deployment levels required for meeting the domestic emissions budget goals discussed in Chapter 2.

Such an assessment is complicated by the considerable uncertainties involved in developing scenarios for how the energy system might evolve in response to particular

---

BOX 3.6
Defining Technical Potential

Our discussion of technical potential refers to the definition developed by NRC (2009a) for the potential "accelerated deployment" options for various energy technologies. "Accelerated" refers to deployment of technologies at a rate that would exceed the reference scenario deployment pace but at a less dramatic rate than an all-out crash effort. These estimates were based on the AEF committee's judgments regarding two factors: (1) the readiness of evolutionary and new technologies for commercial-scale deployment and (2) the pace at which such technologies could be deployed without disruptions associated with a crash effort.

In estimating these factors, the committee considered the maturity of a given technology, together with the availability of the necessary raw materials, human resources, and manufacturing and installation capacity needed to support its production, deployment, and maintenance. In some cases, estimates of the evolution of manufacturing and installation capacity were based on the documented rates of deployments of specific technologies from the past. Note that these estimates do not account for all of the barriers that could practically impede deployment of various technologies (e.g., social resistance and institutional limitations). Thus, the technical potential estimates should be viewed as an upper (optimistic) bound of what deployment level is truly feasible or likely.

TABLE 3.2 Comparison of Projected Requirement (red) and Technical Potential for Deployment (blue) for Various Key Energy Technology Options, for the 167 and 203 Gt CO_2-eq Budget Scenarios

Energy Efficiency (% reduction from ref. case)	2020	2035
Requirement (EMF) for **167** Gt CO_2-eq	2-21	5-33
Requirement (EMF) for **203** Gt CO_2-eq	2-17	4-24
Potential (AEF)	15	30
Nuclear (Twh/y)	**2020**	**2035**
Requirement (EMF) for **167** Gt CO_2-eq	868-1034	1292-2092
Requirement (EMF) for **203** Gt CO_2-eq	869-1014	947-1629
Potential (AEF)	968	1453
Electricity with CCS (Twh/y)	**2020**	**2035**
Requirement (EMF) for **167** Gt CO_2-eq	32-324	233-1593
Requirement (EMF) for **203** Gt CO_2-eq	0-87	0-796
Potential (AEF)	74	1200/1800[a]
Renewable Electricity (nonbiomass) (Twh/y)[B]	**2020**	**2035**
Requirement (EMF) for **167** Gt CO_2-eq	194-688	453-1155
Requirement (EMF) for **203** Gt CO_2-eq	194-593	459-971
Potential (AEF)	811	1454
Biomass Fuels (cellulosic) (mmgal/y)	**2020**	**2035**
Requirement (EMF) for **167** Gt CO_2-eq	17,000-29,000	17,000-33,000
Requirement (EMF) for **203** Gt CO_2-eq	15,000-23,000	17,000-35,000
Potential (AEF)	7,700	26,000

NOTE: AEF estimated technical potential out to 2020 and 2035, and so these years are used as benchmarks for the comparisons with EMF22 estimates.

[a] 1200 is for retrofit or repower of existing plants; 1800 is for new plants.

[b] Estimate is for total renewables, including current capacity and potential new capacity. Potential for 2020 is 10 percent of electricity production, in EIA (2010) as specified in AEF. Does not include hydropower.

GHG emissions-reduction goals. This will depend on many factors, including the types of new policies implemented, the evolution of technology, and the degree to which the barriers particular to individual technology areas can be overcome. As a result, the different models show a wide range of estimates regarding deployment requirements for different technologies. Nonetheless, even taken in a very general sense, comparing the EMF22 and AEF estimates provides significant insights into the feasibility of decarbonizing the energy system.

Both the EMF technology requirements and the AEF technology potentials shown in the table are rough estimates. Taking that uncertainty into account, however, we feel the results are sufficiently robust to make the following observations:

- *For the electricity sector, meeting the 167 Gt CO_2-eq budget would be challenging—requiring that nearly all technologies available to increase efficiency and decarbonize the energy system be deployed at levels close to their full technical potential.* Meeting the 203 Gt CO_2-eq budget is less challenging, but it is nevertheless still very demanding. If CCS can be demonstrated successfully and then deployed widely, this would likely make it feasible for the electricity sector to decarbonize fully. However, CCS has yet to be demonstrated in large-scale utility applications. If it proved to be infeasible, the remaining potential for efficiency, renewables, and nuclear would not be enough to meet electricity needs in 2035. Indeed, if any one of the major categories fails to approach its technical potential, meeting the electricity need would be very difficult.[9]
- *For the transportation sector, meeting the deployment requirements for either budget scenario is particularly difficult.* The technical potential for expanding the use of biomass fuels in transportation appears to be near the low end of what is required. The AEF study shows that, even if we could meet the full technical potential for both vehicle efficiency gains and alternate fuels use, there would still be a need for roughly one-third of the 2035 demand for transportation fuel to be met by oil.[10] This suggests that further displacement of petroleum in the transportation sector will require additional strategies, such as significant deployment of pure or hybrid electric vehicles.
- *The AEF technical potential estimates are based on optimistic assumptions, so falling short of them is quite likely.* AEF does not account for nontechnical (i.e., social or institutional) barriers to deployment; it assumes that the technologies, once adopted, operate at acceptable costs and performance. This provides further impetus to suggest that existing technology options are not likely to be sufficient, and there is an urgent need to enhance R&D aimed a creating new technology options.

---

[9] The Electric Power Research Institute's (EPRI'S) Prism analysis also estimates the technical potential for decarbonizing electric power production. EPRI's estimates are similar to, and in some cases more conservative than, the AEF estimates. Even so, EPRI regards its Prism results to be "very aggressive, but feasible if the proper investments in R&D are made (particularly around demonstration and early deployment)" (personal communication with Bryan Hannegan, Rhode Island)

[10] See Figures 2.4, 2.11, and 2.12 of NRC (2009a) for the data on which this analysis is based.

Inertia of Existing Infrastructure

A second consideration underscoring the need for urgency is that the present energy infrastructure, if left unchanged, will rapidly deplete the GHG budgets discussed in Chapter 2. The reference case in EIA (2010) projects U.S. CO_2-eq emissions to 2035, taking into account the accelerated CAFE standards announced in 2009 as well as the effect of the economic downturn of the past year. It projects annual emissions dropping to a low of 5.7 Gt CO_2-eq in 2013 and then rising to 6.3 Gt CO_2-eq in 2035. Cumulatively from 2012, these emissions amount to 143 Gt CO_2-eq by 2035. This represents 84 percent of the 170 Gt CO_2-eq budget and 72 percent of the 200 Gt CO_2-eq budget, thus substantially truncating the emissions budget for the remaining 15 years until 2050. Some of these emissions could potentially be sequestered through soil and forestry management efforts, but this would slow depletion of the budget by only a few percent. And meanwhile, unchecked GHG emissions from other, nonenergy sources (which were not included in the EIA projections) would further accelerate depletion of the budget.

A similar situation exists globally. As noted in Chapter 2, recent modeling suggests that limiting atmospheric GHG concentrations to 450 ppm CO_2-eq is very difficult, and even holding concentrations to 550 ppm requires aggressive action. Bosetti et al. (2008) examined the costs of delay in a global context and suggested that short-term inaction is a key determinant for the economic costs of ambitious climate policies. That is, an insufficient short-term effort significantly increases the costs of compliance in the long term. Delays in beginning to reduce the U.S. contribution to global GHG emissions would risk further loss of opportunities to control GHG concentrations over the long term.

THE LARGER CONTEXT FOR TECHNOLOGY

Although there are many possible opportunities for limiting GHG emissions, most strategies that the nation could adopt to make large, near-term contributions to reducing emissions center on the deployment of reasonably well-known technologies for energy efficiency and low-carbon energy production. These sorts of technological solutions are the primary focus of both the AEF and EMF22 analyses discussed earlier, and they underlie the case for urgent U.S. action. Chapter 4 focuses on crafting a policy portfolio to accelerate the deployment of these near-term, high-leverage technological opportunities.

Ultimately, however, limiting the magnitude of climate change requires looking beyond just these near-term technological opportunities. One reason for having a broader focus is that we know additional technology choices will ultimately be required. As explained earlier, even if the existing "high-impact" technologies were to meet their full technical potential, they themselves are not likely to be adequate to meet the stringent demands of the emissions budgets discussed in Chapter 2. Our current energy system is largely based on R&D that was done two or more decades ago. Basic research could lead to advanced energy efficiency and supply technologies with greatly improved performance, environmental, and economic characteristics.

Another, perhaps more important, reason to consider a broader suite of strategies is that many barriers inhibit the deployment of even well-known technologies. For example, the adoption of many energy-efficiency technologies and practices requires significant changes in human behavior, lifestyle, and consumer spending practices. New technologies such as CCS are unfamiliar both to the public and to environmental regulators; if experience is any guide, building the required levels of acceptance for such technologies can be an elusive task. Also, inertias in supply chains and interdependent infrastructure systems contribute to slow rates of social and technical change. For these reasons, there is a pressing need for greater understanding of individual and institutional responses to the deployment of new technology.

Thus, technological change (discussed in more detail in Chapter 5) must be set in a larger context of research on how social and behavioral dynamics interact with technology, and how technological changes can interact with broader sustainable development issues. We refer the reader to the report *ACC: Advancing the Science of Climate Change* (NRC, 2010a) for a deeper discussion of these issues and of the profound changes they imply for the scientific enterprise.

KEY CONCLUSIONS AND RECOMMENDATIONS

CO_2 emissions from fossil fuel combustion in the energy system comprise over 80 percent of total U.S. GHG emissions. CO_2 emissions related to energy are driven by economics and demographics and the resulting demand for goods and services, the energy required to produce these goods and services, the efficiency with which energy is produced and used, and the CO_2 emitted by the energy production process.

Numerous opportunities to reduce CO_2 emissions exist, but many of them require time and investment to be developed to the point of deployment, have cost and other implementation constraints, or would have marginal impacts on overall GHG emissions. We conclude that the most substantial opportunities for near-term GHG reductions, using technology that is deployable now or is likely to be deployable soon, include the following:

- Improved efficiency in the use of electricity and fuels, especially in the buildings sector, but also in industry and transport vehicles.
- Substitution of low-GHG-emitting electricity production processes, which may include renewable energy sources, fuel switching to natural gas, nuclear power, and electric power plants equipped to capture and sequester CO_2.
- Displacement of petroleum fuels for transportation with fuels with low or zero (net) GHG emissions.

Meeting the goal of limiting domestic GHG emissions to 170 Gt CO_2-eq by 2050, by relying only on these near-term opportunities, may be technically possible but will be very difficult. Meeting the 200 Gt CO_2-eq goal is more feasible but nevertheless very demanding. In either case, realizing the full potential of known and developing technologies will require reducing many existing barriers to deployment; therefore, it is likely these technologies will fall short of their technical potential.

This underscores the crucial need to strongly support R&D aimed at bringing new technological options into the mix (discussed further in Chapters 4 and 5). Meeting the 2050 budget goal requires that these new technologies be available by the 2020-2030 time period. To create the necessary innovations in time for deployment means moving research along very rapidly.

Some important opportunities exist to control non-CO_2 GHGs including CH_4, N_2O, long-lived fluorinated GHGs, and short-lived pollutants such as ozone precursors and black carbon aerosols. Opportunities also exist to enhance biological uptake and sequestration of CO_2 through afforestation and soil management practices. These opportunities are worth pursuing, especially as part of a near-term strategy, but they are not large enough to allow the United States to avoid falling short in reducing emissions from fossil fuel energy sources.

Our nation's existing energy system, if left unchanged, will rapidly consume the emissions budgets suggested in Chapter 2 (especially the more stringent 170 Gt CO_2-eq budget). Delay in reforming the energy system would thus make a challenging goal essentially unattainable.

Because of this compelling case for urgency, we conclude that action is needed: to accelerate the deployment of technologies that offer significant near-term GHG emissions-reduction opportunities; to accelerate the retirement or retrofit of existing high-emitting infrastructure; and to aggressively promote research into the development and deployment of new, low GHG-emitting technologies.

CHAPTER FOUR

Crafting a Portfolio of Climate Change Limiting Policies

Reducing the threat of climate change will require providing the right incentives for behaviors and investments that drive a transition to a low-greenhouse-gas (GHG) emissions economy. One means of doing so is to create price signals that reflect the costs associated with GHG emissions. The pricing instruments most commonly considered, carbon taxes[1] and cap-and-trade programs, both create incentives that are compatible with cost-effective reduction of GHG emissions.[2] It is our view that a pricing policy, properly designed, is essential for creating broad incentives for emissions reductions; but evidence suggests that pricing alone will not be sufficient to achieve the necessary emission reductions (Fischer and Newell, 2008; Goulder and Parry, 2008), and carefully tailored complementary policies will be needed to address shortcomings in a pricing system.

In this chapter, we first describe the common design features of carbon-pricing schemes, including the scope of gases and emission sources covered, the points of control, how revenues can be used, and how pricing can be enforced. We then discuss how certain design choices can dilute or undermine the effectiveness of a pricing strategy, and why even a well-designed pricing strategy will have limitations that restrict the timing and scope of its effectiveness. We then identify the crucial targets of opportunity for future reduction of GHG emissions and identify a series of possible complementary policies targeted at those opportunities. Finally, we discuss the challenges of integrating these different policies into a cohesive whole. This chapter focuses primarily on national-level policy responses. The important role for state- and local-level policy responses (in relation to federal policy) is discussed in Chapter 7.

---

[1] As noted earlier, we treat the terms "carbon price" and "carbon tax" as synonymous with the more general terms "GHG price" and "GHG tax," as they are in most instances applied to multiple gases.

[2] As discussed later in this chapter, it is possible to use both instruments simultaneously in a hybrid system.

PRICING STRATEGY DESIGN FEATURES

Carbon taxes and cap-and-trade polices are usually discussed in terms of their differences, but many of the same design questions need to be resolved for each. Some of the key questions are discussed below.

The Scope of Coverage

Cap-and-trade policies that cover only CO_2 may be administratively convenient, but they do not represent the best long-term solution. The Kyoto Protocol, for example, identifies six GHGs, which can be included in a single pricing system by translating them into CO_2 equivalents. In practice, this is accomplished using global warming potentials (GWPs), defined as the cumulative radiative forcing effects of a unit mass of gas relative to CO_2 over a specified time horizon (commonly 100 years). Including multiple gases under a single cap has the advantage of significantly reducing the cost of reaching a specific concentration target (Reilly et al., 1999; Weyant et al., 2006). Disadvantages include controversies over whether GWPs are an appropriate metric to account for the differing impacts among GHGs, and the fact that some types of GHG emissions (e.g., those stemming from land-use and agricultural practices) are quite difficult to monitor.

For maximizing GHG emissions reductions at minimum cost, more universal coverage is better. Yet no existing program involves universal coverage of GHG sources. As two key examples, the Regional Greenhouse Gas Initiative (RGGI) in the Northeast covers only large power generators, and the European Union Emissions Trading Scheme (EU ETS) covers only power generators and combustion installations, production and processing of ferrous metals, pulp, and paper, and some mineral industries such as cement (and for each sector, only facilities over a specified size are typically covered); in addition, aviation will be covered starting in 2012.

Extending coverage beyond these typical sectors does present challenges. Omitting non-CO_2 GHGs and emissions and sequestration in the agricultural and forestry sectors is generally motivated by concerns about political feasibility, impressions that sequestration is a means of avoiding needed emissions cuts, and uncertainties in the magnitudes of potential reductions from particular sectors. In addition, smaller sources may face unreasonably high transaction costs in complying with a one-size-fits-all program. Below we discuss how these challenges can be addressed through the proper design of pricing mechanisms and how offsets can be used to address some sources not directly covered in a cap-and-trade policy.

Targeting the Control Responsibility

In general terms, the choices for applying controls (i.e., who is assigned the cap) include upstream targeting, downstream targeting, or a hybrid involving some combination of the two. In an upstream point of regulation, allowances are surrendered at the point of extraction, production, import, processing, or distribution of substances that (when used or combusted) result in GHG emissions. This approach was originally developed with fossil fuels in mind, but it could be extended to other gases. A downstream point of regulation would focus control on the point of emission into the atmosphere (power plants, cars, etc.).

An upstream approach controls emissions indirectly rather than directly. For example, energy suppliers would either have to employ technologies to reduce the carbon in their fuels or buy allowances to cover what remains. Since fuels with high-carbon content would need relatively more allowances per BTU of energy, they would experience a relative increase in their cost—an increase that would be passed forward to consumers. This higher cost of energy in general would promote greater investments in energy efficiency, and the relative price increase for high-carbon fuels would promote some substitution of fuels with lower carbon content.

Because it involves monitoring fewer parties, an upstream approach would likely have lower administrative costs. However, it would necessitate a system for rebating fees for feedstocks that are not combusted and therefore do not become GHG emissions (such as oil used for lubrication) and for combustion gases that are captured and sequestered rather than emitted. Like so many other design choices, the point of regulation is not necessarily an either/or choice. Hybrid strategies, involving upstream control of some sources and downstream control of others, are also possible.

Allocating Entitlements

Both tax and cap-and-trade policies control access to the use of the atmosphere as a repository for emitted GHGs. When this access is limited, the access rights become very valuable, and the initial allocation of these rights can advantage certain groups. To whom, and under what terms, should this value accrue?

Both tax systems and auctioned cap-and-trade systems force users to pay for that access. This approach generates revenue—in the case of GHG control, a considerable amount of revenue.[3] The implicit logic behind this approach is that the atmosphere belongs to all the people and the wealth created by allocating scarce access rights should be returned to the people or used for public purposes. This is the approach taken by the RGGI program, in which all participating states are auctioning at least the majority of allowances. The alternative is to gift some or all of the allowances to parties based upon some eligibility criteria (e.g., allocations to firms with best practices in an industry, actual historic emissions, or even allocations targeted directly to households).

There can be strong political motivations to give away (gift) emissions allowances, as this offers a way for policy makers to gain support from particular industries or constituencies who would otherwise strongly oppose a carbon pricing system. Research strongly suggests, however, that use of revenue-raising instruments (either taxes or auctioned emissions allowances) is more economically efficient than gifting.[4] This efficiency advantage results from a balance between two effects: a "tax-interaction" effect that intensifies preexisting market distortions and thus reduces general welfare, and a "revenue recycling" effect that mitigates preexisting market distortions and thus increases general welfare (Goulder, 1997). When the second effect is larger than the first, it can produce a "double dividend"—environmental benefits, and the welfare gained from revenue recycling.

Distributional biases can also occur with revenue-raising instruments (Parry et al., 2006). As discussed further in Chapter 6, the cost burden from a gifted cap-and-trade system (where the allowances are given directly to firms) is strongly regressive; that is, it is borne disproportionately by lower-income households (Chamberlain, 2009; Dinan, 2009).[5] This is due in part to the inherently regressive nature of the policy, and in part to the fact that gifting to firms allocates the value to the shareholders of the gifted companies (who are generally in higher income brackets). Gifting allowances directly to lower-income households diminishes the regressivity.

The experience in the EU ETS has enriched our understanding of the dynamics of gifting allowances to firms. Empirical evidence has demonstrated that, in deregulated electricity markets (mainly the United Kingdom, the Netherlands, Germany, and the Nordic countries), allowances that were gifted to electricity generators allowed those

---

[3] At $30 per ton, current emission rates of ~7,077 million metric tons of CO_2-eq per year (supplied by the Energy Information Agency) would yield annual revenue of $212.3 billion.

[4] This analysis compares efficient policies. The comparison may not hold if the polices in question are riddled with exemptions or exceptions.

[5] These analyses are generally based on an implicit assumption that the United States alone is taking mitigation measures. A broader global market would affect energy prices internationally, which would in turn influence the distributional burden on the poor.

parties to capture the full value of these allowances without incurring any cost, resulting in what has become widely perceived as "windfall profits" (Sijm et al., 2006).

The Congressional Budget Office (CBO, 2009) estimated that, in a scenario where emissions were reduced by 15 percent and all of the allowances were distributed free of charge to producers in the oil, natural gas, and coal sectors, the value of the allowances would be 10 times the combined profits of those producers. The windfall gains received as a result of the free allocation would far outweigh the loss in sales that might be experienced when consumers cut back on use of fossil fuels (Dinan, 2009). This finding has the important implication that, even if it is deemed politically necessary to gift some allowances (for instance, to reduce the trade vulnerability of certain energy-intensive industries), it can be accomplished with a relatively small proportion of the total value.

Using Funds from Taxes or Auctions

The distribution of revenue from auctioned allowances or carbon taxes can, in principle, enhance policy efficiency or help reduce the regressive financial burden of emissions-reduction efforts. Those benefits, however, depend upon what is done with the revenue. Evidence presented by the CBO suggests that rebating the funds back to households (on a per capita lump-sum basis) converts the regressive policy associated with gifting allowances to firms into a progressive policy. That evidence also suggests that a rebate to households is more progressive than reducing the payroll tax and much more progressive than reducing the corporate income tax (Dinan, 2009). Focusing exclusively on distributional goals and returning all revenue to households requires a trade-off with the efficiency gains from reducing distortionary taxes (Dinan and Rogers, 2002). Some recent work, however, suggests it is possible to do both while still protecting vulnerable industries. Goulder et al. (2009) suggest, for example, that vulnerable industries could be protected by gifting 15 percent or less of the allowances and auctioning the rest to raise revenue for pursuing the distributional and efficiency goals.

Competition from other uses of tax or allowance revenues is inevitable. To name a few:

- Energy-intensive, trade-vulnerable firms may seek financial rebates as protection against competition from foreign firms that are not subject to control of GHG emissions.
- States running their own cap-and-trade programs will seek to replace funds lost if a federal preemption results in the demise of these programs (and in the

funding dedicated to promoting energy efficiency and renewable resources that states have raised from auctions).

- Negotiators seeking to bring developing countries into a binding international agreement will be looking for funds to facilitate the transition.
- Federal departments charged with promoting new technologies or strategies will be looking for funds for research and development (R&D), for startup incentives, and for demonstration projects.
- Funds from GHG control are tempting to use as incentives as Congress tries to build coalitions of legislators to ensure the passage of climate change legislation.
- Other public issues such as health care may seek sources of funding, based on the rationale that climate change does affect health.

The Impact of Design on Allowance Prices

Estimating the costs and benefits of a program to limit GHG emissions is difficult because it depends on many factors that are unknown or uncertain at the time the estimates are produced and because it depends on specific characteristics and assumptions in the models being used to produce the estimates (e.g., the degree of aggregation or the handling of technical change). Estimates from the literature may vary significantly simply because the models used to derive the estimates have differing assumptions about underlying policy packages.

As discussed later, future allowance prices can be lowered by implementing complementary policies, for instance, policies that lower the demand for energy through efficiency measures, that increase low- or zero-carbon energy supplies, that allow offset credits for reductions not covered by the cap, and that promote the early introduction of carbon capture and storage (CCS). Lower allowance prices can have the advantage of lowering the financial burden on businesses and households,[6] and limiting the potential competitive disadvantages and resulting emissions leakage if other countries do not follow suit. However, the disadvantage is that lower allowance prices may delay investment in more expensive, low-emitting, new technologies simply because the value of the emissions saved is too low to justify the investment.

---

[6] As discussed later, lower allowance prices may not always reduce the burden on firms and households; if the costs associated with complementary policies are high enough, they can more than offset the advantages from lower allowance prices.

The Role for Offsets and Offset Tax Credits

Offset credits reflect emissions reductions for sources that are not covered by the cap or not included in the base of a GHG tax but which can be credited against the cap or tax base by the acquiring party. Offsets (or offset tax credits) can perform several useful roles. First, by increasing the number of emissions-reduction opportunities, they lower the cost of compliance. Second, they extend the reach of the tax or cap by providing incentives for reducing sources that are not directly covered.[7] Third, because offset credits separate the source of financing reductions from the source that actually provides the reduction, they can help secure some reductions using capital that, for affordability reasons, might not otherwise be mobilized for this purpose. Both extending the reach of the cap and offering financing may be crucial for ensuring that meaningful reductions take place in developing countries.

Some emissions sources are difficult to include directly within a pricing system. For example, fugitive emissions (arising from leaks during the processing, transmission, and/or transportation of GHGs) are very difficult to monitor and, hence, enforcement based on actual emissions would be very difficult. In these cases, offset credits can be used to secure reductions from specific projects where the reductions can be monitored and validated (i.e., projects that are capable of securing certifiable reductions). When certified, these credits can then be used by acquiring entities as one of the means of meeting their cap obligation or reducing their tax base.

Potentially the most serious problem facing offset certification is demonstrating compliance with the "additionality" requirement. An emissions reduction is considered "additional" if human-caused emissions of GHGs from that source are reduced below what would have occurred in the absence of the offset activity. In practice, that is not a trivial determination, and it often requires consideration of factors such as financial motivation and regulatory context for an activity. There is an inherent tension between the need to hold transaction costs down and the need to provide assurance that the credited reductions are real and additional. Putting considerable effort into establishing a baseline and verifying reductions is important but costly. As the transaction cost associated with certifying offset projects rises, their profitability, and hence their supply, falls. This was the case in the early U.S. Emissions Trading program for SO_x during the 1970s and 1980s (Dudek and Palmisano, 1988; Hahn and Hester, 1989).

Internationally, the Clean Development Mechanism (CDM) under the Kyoto Protocol is the largest forum for the development and use of offsets, known in that program

---

[7] Current examples from RGGI include credits for reducing methane from landfills or for the additional carbon absorption resulting from a reforestation effort.

as certified emission reductions (CERs). The CDM provides a useful example of how an offset program can work in practice. Despite continued concern over transaction costs (Michaelowa and Jotzo, 2005), the CDM has stimulated a considerable amount of investment. As of April 2009, it had registered some 1,596 projects resulting in over 280 million tons of emissions reductions as CER credits. By the end of 2012 it expects to have issued CERs of more than 1.5 billion tons.[8]

The CDM program also illustrates some sources of controversy associated with offsets, including the types of projects being certified (an alleged overemphasis on non-CO_2 gases), the skewed regional distribution of CER activity (with China, India, South Korea, and Brazil creating more than 60 percent of generated credits), and the amount of subsidy being granted (with actual emissions-reduction costs being well below the price received for a CER) (Wara, 2007). Another more global concern is that the CDM creates adverse incentives for host countries to pursue reductions on their own (i.e., developing counties may well hesitate to undertake projects on their own, as long as they can get someone else to pay for them through CDM) (Hall et al., 2008).

Controversies about the validity of CDM credits outside the range of domestic monitoring have led to resistance to the blanket use of nondomestic offsets (Wara, 2007). Yet the large potential impact of these offsets on allowance prices and compliance costs has created pressure for some middle ground, where international offsets are used, but only in a controlled environment where their validity can be ensured. Several types of approaches are available in this regard.

One approach for ensuring that actual domestic reductions are sufficiently high is to restrict the use of offsets (either domestic or foreign) to some stipulated percentage of the total required allowances.[9] Disadvantages of this approach are that it raises compliance costs and fails to distinguish between high- and low-quality offsets. A second approach is based on distinguishing between offset types; that is, programs are open to high-quality offsets, but not to low-quality offsets. A U.S. program following this approach would need to establish eligibility criteria to identify which offset types are acceptable and to not allow those that do not meet the criteria (Hall et al., 2008). A third approach is to discount the amount of emissions reduction per offset (or the allowance price) to provide a margin of safety against uncertainty in the magnitude of the reductions that may result from this offset project. Discounting can specifically

---

[8] The official data can be found at *http://cdm.unfccc.int/index.html* (accessed April 28, 2009).

[9] In the RGGI, for example, CO_2 offset allowances may be used to satisfy only 3.3 percent of a source's total compliance obligation during a control period, though this may be expanded to 5 percent and 10 percent if certain CO_2 allowance price thresholds are reached. Although the intention to allow limited use of CDM credits has been stated, to date the specific rules for allowing those credits remain unspecified.

address concerns such as permanence, additionality, and leakage (Kim, 2004; Smith et al., 2007). A fourth possible approach is to allow offsets for specific countries that fulfill monitoring and certification requirements but to explicitly phase out those offset credits over time, to prevent the offset opportunity from creating incentives against acceptance of an emissions cap by the host countries. See also the discussion of offsets in Chapter 2, which raises the idea that a heavy reliance on international offsets could possibly result in needs for additional compensation to the seller countries.

Offsets can thus play a useful role both in lowering costs and in involving international participants, but it must be a carefully circumscribed role with effective oversight or else the liberal use of offsets could reduce the likelihood that GHG reduction goals will be met. Putting considerable effort into establishing an appropriate baseline and verifying reductions is costly; this cost creates a tension between the desire to increase the supply of offsets and the desire to ensure the environmental integrity of the program. Furthermore, using widespread offsets to lower the GHG price can delay the development and use of some new low-emission technologies that can only be justified at higher prices. Finally, a number of practical implementation concerns raised by offsets are described in Box 4.1.

COMPARING TAXES WITH CAP AND TRADE

As discussed above, many aspects of designing polices to put a price on GHGs are similar for both a tax and a cap-and-trade policy. Those similarities, however, should not obscure the important differences that exist as well, as summarized below.

Linking to the Existing System

The United States has considerable experience with cap-and-trade programs that goes back to the mid-1970s, including the highly successful sulfur allowance program (Ellerman, 2000; Tietenberg, 2006). It does not have similar experience with using taxation to control pollution, but it does have considerable experience with (and infrastructure for) levying taxes in general.

Generally the targets for environmental policy are stated in quantity terms (concentration or aggregate emissions limits). Meeting quantity limits is easier with a cap-and-trade policy than with a tax policy, simply because the cap can be set equal to the aggregate emissions goal, but the price that would achieve that goal is not known in advance and can only be approximated.

BOX 4.1
Offsets: Practical Implementation Concerns

In order to be credible, offsets must be real, additional, quantifiable, verifiable, transparent, and enforceable. Guaranteeing these properties requires establishing an administrative system with several key elements, discussed below.

Certification standards. Offsets can be reviewed on a case-by-case basis (generally the approach taken for CDM projects under the Kyoto Protocol) or they can be subject to uniform performance standards by sector (for example, the Climate Action Reserve has protocols for sectors such as urban forestry, livestock, and landfills). The best strategy may be a hybrid approach that relies on the development of standardized protocols but maintains a significant amount of regulatory oversight of individual projects. This is the approach California is considering in the design of its offset program. This approach is also recommended by the Offset Quality Initiative, a joint program of nonprofits involved in the development of climate change limiting policy.

Certification process. Who should be responsible for certifying or verifying the credibility of offsets? The CDM relies on independent third-party verifiers called Designated Operational Entities; the RGGI also uses independent verifiers, while the California Climate Action Registry has adopted its own protocols for verifying offsets. Alternatively, the certification could be done by a federal agency such as the Environmental Protection Agency (EPA). An advantage of relying on independent verifiers is that many have already developed significant expertise; however, a disadvantage is that the government has less control over independent entities engaged in certification.

Enforcement. Because the establishment of a well-functioning program to oversee offsets will be complicated and highly technical, Congress may need to delegate authority to decide precisely how to design an offset program to an administrative agency such as the EPA. This agency will need the authority to investigate, subpoena records from, and penalize entities that violate the rules of the offset program, including any third-party independent verifiers, developers of projects used for emissions reduction, and regulated entities seeking to use offsets to meet their regulatory obligations. In addition, Congress should consider including a citizen suit provision within cap-and-trade or tax legislation, allowing individuals to enforce the offset provisions against violators.

Staffing and financing. The verification of offsets is likely to be a labor-intensive process. It is vital that the regulatory authority have the personnel necessary to ensure the integrity of the program. Without adequate staff, the entire credibility of a cap-and-trade program that contains offsets could be undermined. Congress may wish to consider imposing a fee on applicants for offset project approval sufficient to cover the administrative costs of oversight.

While a few carbon tax systems exist in Europe, most existing GHG control programs (e.g., the Kyoto Protocol, the EU ETS, and the RGGI) are based on a form of cap-and-trade policy. A U.S. national cap-and-trade program could integrate with existing systems (eventually permitting allowances to be traded between programs).[10] Assertions that a tax system could not be similarly integrated, however, are not merited. For example, in a carbon tax system, Certified Emission Reductions from the CDM could easily be authorized to serve as tax offsets, and U.S. firms could sell offsets to international buyers. Trading allowances, however, would have no counterpart in a tax system.

Supporters of cap and trade point out that the existence of an active carbon market could serve as a considerable lure for developing countries. These countries would almost surely be net sellers in a global carbon market and could expect to earn substantial profits from abating emissions and selling allowances. Meanwhile, because advanced economies like the EU and the United States can set the terms of access to their own markets, they would have considerable leverage to persuade those other countries to take on binding emissions targets.[11] An emissions tax provides neither such an incentive nor such leverage (Keohane, 2009).

Policy Stability

One desirable aspect of any GHG pricing strategy is a stable policy platform designed to reduce regulatory uncertainty associated with energy investments. In principle, both a tax and a cap-and-trade mechanism would provide policy stability, but the form differs.[12] While a carbon tax fixes the price of CO_2 emissions and allows the quantity of emissions to adjust, a cap-and-trade system fixes the quantity of aggregate emissions and allows the allowance price to adjust. In practical terms, this means a cap-and-trade policy provides more certainty that the GHG reduction goal would be met, but it provides less certainty about the costs. Conversely, a tax policy provides more inherent certainty about cost, but less certainty about the resulting emissions levels. The uncertainty over emissions reductions associated with a tax approach can be lessened using the adaptive design features discussed below; however, to the ex-

---

[10] Integration is not trivial. For an expanded exploration of the linkage possibilities offered by cap and trade see, Jaffe and Stavins (2008).

[11] Admittedly, these arrangements would probably supersede the CDM, and governments of some developing countries might be reluctant to see such a change. However, the volume of credits under such a system would be much greater than in the CDM, and the great benefits to be received by developing countries would provide incentives for them to accept more credible monitoring and compliance institutions.

[12] In practice, initially determined tax rates or caps may be changed by subsequent legislative action, thereby undermining the stability on which this comparison depends.

tent changes are invoked by that process, the advantage of price stability will be lost. Similarly, more control can be exerted over prices in a cap-and-trade system, but any such intervention diminishes the degree of certainty about the resulting emissions.

Price Volatility and Cost Containment

A tax system fixes prices, and so in the absence of any intervention to change them, price volatility is not an issue. That is not the case with cap and trade, either in principle or in practice. Several cap-and-trade programs (e.g., EU ETS, RECLAIM,[13] and the U.S. SO_x emissions trading program) have experienced price volatility. Some of this volatility resulted from correctable design defects such as overallocation of the allowances in the first phase of the program, lack of up-to-date information about emissions levels, and a failure to permit allowances in the first phase to be banked for use in the second phase. Although such defects are correctable, that does not mean the issue of price volatility is easily dismissed.

Price volatility can potentially be addressed by coupling a "price collar" consisting of a price floor and ceiling, with an allowance reserve. A price floor (which has been adopted by the RGGI program) would help alleviate investment problems and revenue shortfalls resulting from allowance prices falling to unacceptably low levels. A price ceiling would permit additional allowances to be purchased at a predetermined price set sufficiently high that it would become a binding constraint only if allowance prices exhibited drastic spikes. To prevent these purchases from breaking the cap, they would come from an allowance reserve, established from allowances set aside for this purpose from earlier years, from an expansion in the availability of domestic or international offsets, or perhaps from allowances borrowed from future allocations (Burtraw et al., 2009; Jacoby and Ellerman, 2004; Murray et al., 2009a; Pizer, 2002). One disadvantage of a price ceiling is that it can undermine incentives for developing new technologies that would be justified only by prices higher than the ceiling.

Temporal Flexibility

Both cap-and-trade and tax systems offer options for temporal flexibility.[14] For cap and trade, this flexibility is achieved by allowing *banking* (holding an allowance beyond

---

[13] RECLAIM stands for the Regional Clean Air Incentives Market program of the CA South Coast Air Quality Management District.

[14] Murray et al. (2009a) suggest that in a dynamic setting the cap-and-trade policy may have one distinct advantage—an advantage that arises from the unique combination of intertemporal flexibility and foresight

its designated year for later use), *borrowing* (using an allowance before its designated date), or both.[15] The economic case for banking and borrowing is based on the additional flexibility it allows sources in the timing of their abatement investments. Flexibility in timing is important because the optimal timing for installing new abatement equipment, or changing the production process to reduce emissions, can vary widely across firms.

Price considerations also argue for temporal flexibility. If all firms were forced to adopt new technologies at exactly the same time, the concentration of demand at a single point in time (as opposed to spreading it out) would raise prices for the equipment, as well as for the other resources (such as skilled labor) necessary for its installation.

Banking also has the potential to reduce price volatility. Storing permits for unanticipated outcomes (such as an unexpectedly high production level, which triggers higher-than-expected emissions) can reduce future uncertainty considerably. Because stored permits can be used to achieve compliance during tight times, they provide a safety margin against unexpected contingencies.

Empirical evidence and experience with existing programs support the idea of allowing banking in any cap-and-trade program (Tietenberg, 2006). The evidence for borrowing is weaker, because it is an uncommon feature in existing programs. A regime that allows unlimited borrowing raises at least one potential concern: that it could reduce flexibility to tighten the cap in cases where new scientific information suggests that doing so is necessary.

Administrative Ease

In general, a cap-and-trade policy and a tax policy require many of the same administrative functions (e.g., defining the goals, monitoring emissions, and ensuring compli-

---

afforded by markets. Through dynamic market arbitrage, the cap-and-trade system allows any expectations about future benefits, costs, or target modification to be transmitted to markets (and market prices) today. The tax instrument, in contrast, does not automatically respond to changes in expectations. With a tax instrument, even if firms correctly anticipate a higher marginal cost or tax in the future, they cannot arbitrage against this outcome by overcomplying now and banking allowances for use in the future. While taxes (like the cap) can of course be adjusted over time, inefficiently high or low levels of abatement and costs will be experienced during the periods between adjustments.

[15] The Corporate Average Fuel Economy (CAFE) standards program also allowed "trading" in the form of carry-forwards and carry-backs, though only within a single company, and the carry-forwards and carry-backs had a limited lifetime. Thus, when gasoline prices fell in the mid-1980s and the domestic auto manufacturers went from exceeding the CAFE standards to falling short, a significant share of the credit carry-forwards they had accumulated expired before they could be used.

ance). The use of an offset market increases this burden considerably, but this additional burden would be borne by either approach as long as offsets were included. The cap-and-trade policy, however, does have one additional administrative burden: it must create and administer an allowance market. This responsibility includes the establishment of allowance registries to keep track of all transactions and provision of a market where buyers and sellers can meet. These are both familiar functions with strong precedents provided by the sulfur allowance programs and the RGGI, but they do add an additional administrative requirement.

The Role of Uncertainty

A crucial feature of climate change is uncertainty about the costs of emissions abatement— especially for technologies yet to be deployed or developed. As is well established in the economics literature, when marginal costs are uncertain, the relative efficiency of a price instrument versus a quantity instrument (i.e., tax versus cap and trade) depends on the relative slopes of the marginal benefit and marginal cost functions (Weitzman, 1974). This insight has implications for policy instrument choice. On efficiency grounds, a cap-and-trade policy will be preferable when the marginal benefits slope is steep relative to marginal costs. The flatter the marginal benefits slope, the stronger the preference for a tax.

The prevailing view in the economics literature is that the marginal benefits of reducing GHG emissions are likely to be flat, since the damages from climate change are driven by the accumulated concentration of GHGs (e.g., Nordhaus, 2008). This view has been supported by analyses that find a strong preference for a price instrument (Hoel and Karp, 2001, 2002; Newell and Pizer, 2003). The implication is that when marginal costs are quite sensitive to the level of emissions reduction but the damages from climate change are not, a carbon tax is preferred on efficiency grounds.

The conventional wisdom, however, assumes that the effects of climate change increase steadily as a function of atmospheric concentrations of GHGs (as discussed by Keohane, 2009). In fact, growing scientific evidence suggests that climatic responses to temperature increases are highly nonlinear and may well be characterized by thresholds or abrupt changes (see *ACC: Advancing the Science of Climate Change* [NRC, 2010a]). Incorporating these threshold effects in damage estimates leads to a greater sensitivity of damages to the level of emissions reduction, shifting the preference toward cap and trade. At the very least, the rationale for an emissions tax in the presence of cost uncertainty is on much less solid ground than is usually assumed in conceptual models. More research on this point is needed before one policy instrument can be said to dominate the other on these grounds.

Policy Durability

The United States needs a sustainable policy for limiting GHG emissions—one that can last for many decades. It will not be sufficient to *enact* a well-designed carbon-pricing policy; it will be necessary as well to *maintain* that policy despite predictable political pressures to relax it or to create exceptions that undermine policy objectives. This reality is one instance of a generic sustainability problem, in which reforms in the public interest can sometimes be enacted, aided by high levels of media attention. But it is often difficult, when public attention wanes, to prevent those reforms from being undermined or distorted by subsequent special-interest politics. In Chapter 8, we examine historical examples of policy reforms that either proved durable or did not. The lessons to be learned from these examples are that reforms are sustainable when the major players have interests in their continuation; that reforms should be designed to provide *incentives* to firms to make investments that are contingent on maintenance of the programs; and that, in general, incentives should be aligned in a way that is self-reinforcing.

Both tax and cap-and-trade systems would have some participants seeking to weaken or repeal the system over time. However, a cap-and-trade system allocating allowances with a market value provides clearer incentives for GHG emitters to insist on maintaining the policy framework. Those buying, banking, and selling allowances (as well as entities involved in the infrastructure of the carbon market—brokerage houses, registries, etc.)—are likely to insist on preserving a stable market. Such effects were evident in the experience of the U.S. SO_x cap-and-trade market (which was reinforced by the fact that the EPA managed the trading system well and that manipulation of the system was difficult).

Policy Adaptability

Although policy must be stable over time, it must also be flexible enough to incorporate new information. Modifying policy frameworks in the face of new information has precedent, for instance, in the Montreal Protocol: When better scientific information confirmed that more stringent targets were needed, the treaty targets were changed. Yet system modification, if not done carefully, has the potential to undermine incentives that provide the system's foundation. The stable, predictable prices established by a tax system provide the market security that investors depend on when making long-term investments. Similarly, cap-and-trade systems depend on allowance holders having secure ownership rights to the allowances. When a new understanding of the science makes adjustments in the tax rates or the caps necessary, that security can be jeopardized.

The desire to allow change can be made compatible with the desire to preserve sufficient security for investors by using an adaptive management system and flexible policy instruments (Arvai et al., 2006). In an adaptive management system, initial programs are designed in a way that allows learning about the impacts of the program; that knowledge is then used to improve subsequent programs. This is especially useful for issues such as climate change, which require many combinations of policies with complex patterns of interaction.

A key requirement is to have a transparent process for dealing with evolution of the system over time. This would include a specification of trigger points for initiating investigation of the need for modification and for invoking the policy change itself. The outcomes of policies should be reviewed on a periodic schedule, and the process for deciding whether (and how) policies are changed, and for enacting needed modifications, should be made transparent to interested parties from the outset.

Suppose, for example, that new scientific evidence suggests the need for further emissions reductions. For a tax system, this means that future tax rates need to be raised. This can be handled by specifying in advance the percent increase in rate to occur as a function of the new emissions reductions needed. Existing estimates of price elasticity in emissions reductions could be used to specify those relationships, and those estimates can be improved over time as more experience is gained. For a cap-and-trade system, a need for additional emissions reductions would simply be met by defining lower caps. The authorized level of emissions for banked allowances and offsets should be unaffected by a change in the cap. Government should not confiscate banked credits not currently in use, because this destroys the incentive to create excess entitlements.[16] In general, gifted allowances should be defined as a percentage of the cap, not as a specific number of tons (a system common in fisheries with transferable catch quotas), because this allows the cap to be changed without forcing the government to buy back the resulting surplus allowances (Ostrom, 2002).

The policy mix can also evolve over time. Adjustments might be necessary, for example, because interaction effects among different policies turn out to be more

---

[16] While this proposition may appear to be self-evident, in fact confiscation of banked credits took place in the United States in the early years of the SO_2 emissions trading program and in the CAFE Program.

important than originally anticipated. Midcourse corrections can incorporate the information gained from experience. If current policies are not achieving the desired level of investments in technological innovation or of emissions reductions, then additional policies can be added to the mix, policies can be phased out, or the design of individual policies can be refined. In each case, it is important to have the evaluation and adjustment process spelled out in advance so that all participants understand the ground rules.

Hybrid Options

Many possible forms of policy hybrids exist, some of which are dealt with elsewhere in this chapter. Here we focus on options for hybrid systems that blend cap and trade with tax policies, which can potentially exercise some degree of control over both emissions quantity and emissions price. The use of ceiling and floor prices in a cap-and-trade policy, discussed earlier, could be viewed as one example of this type of hybrid system. Another, quite different, option is the use of cap and trade for some sectors (e.g., electric utilities) and the use of taxes for others (e.g., transportation, and heating fuels).

Yet another hybrid option involves combining a minimum tax on GHG emissions with a cap for trading. This approach was used in the U.S. approach to complying with the Montreal Protocol, where low tax rates were coupled with (gifted) production and consumption quotas; over time the tax rate was raised sufficiently high that the demand for allowances was driven to zero, making taxes the de facto sole policy. When this sort of cap-and-trade system gifts all allowances, it offers the advantage of raising revenue, which can be used in lowering the regressive impacts of the policy, in promoting complementary policies, or for other useful purposes. When allowances are auctioned, however, this advantage disappears. Although the minimum tax lowers the auction price, it does not affect the amount of revenue flowing to the government, because the government is getting all the auction revenue anyway. Another advantage of such a system is that it may provide a means of transitioning to a tax (if that is the desired outcome) in cases when near-term political realities may constrain that option.

Taxes Versus Cap and Trade: Summary

In summary, the panel strongly supports a carbon-pricing system to provide economic incentives for limiting emissions. Either a cap-and-trade policy or a tax policy could work effectively, and each offers advantages and disadvantages, as described above.

A tax system offers some unique potential advantages such as more price certainty and, for sectors with numerous small emitters, easier administration. Cap and trade, however, has been the option chosen in all recent U.S. national policy proposals, and we see no strong compelling reason to argue that this approach should be abandoned in favor of a taxation system. In fact, we suggest that a cap-and-trade policy would generally be more consistent with the strategies proposed in this report, for several reasons:

- With regard to economic efficiency, in a tax system, concerns about equity and appeasing certain constituencies are handled by tax exemptions, which generally undermine economic efficiency because exempted emissions are uncontrolled. Under cap and trade, such concerns are handled by gifting allowances, which at a first approximation does not undermine efficiency because one retains control over the level of aggregate emissions.
- A cap-and-trade system provides greater certainty about quantities of emissions to be reduced, and thus it is directly compatible with a cumulative emissions budget. In fact, the emissions budget *is* the cumulative cap, and annual increments of emissions reductions would have to conform to this cap.
- As discussed earlier, a cap-and-trade system that allocates allowances with a market value provides clearer incentives for GHG emitters to insist on maintaining a stable policy framework, thus advancing the goal of policy durability
- Most countries that have thus far set reduction targets have relied on cap and trade as the main policy mechanism.[17] A cap-and-trade system also creates incentives for low- and middle-income countries to institute their own country-wide or sectoral caps to derive revenue from selling emissions allowances in the U.S. market. A U.S. cap-and-trade system would thus be consistent with, and help reinforce, the developing international regime.

COMPLEMENTARY OPTIONS FOR THE POLICY PORTFOLIO

As discussed in the previous section, theory suggests that imposing a sufficiently high market price on GHGs will produce the greatest incentive for innovation and lead to the pursuit of lowest-cost means of emissions reduction across economic sectors (Fischer and Newell, 2008; Tietenberg, 2006). In practice, however, pricing alone is not likely to be sufficient because of two interrelated problems. First, the initial design of the pricing mechanism is likely to have shortcomings that will require remedy over time. Second, market barriers exist that inhibit response to price signals.

---

[17] A few European countries have proposed or instituted carbon tax systems, but generally this is supplementary to the EU-wide cap-and-trade scheme.

Rationale for Complementary Policies

Political Realities Will Dictate the Initial Design of a GHG Pricing Scheme

The political process may require some pragmatic considerations regarding the timing, coverage, and structure of a pricing system that will make it less than ideal in terms of efficiency and effectiveness. As an example, to soften the impact of pricing on sectors and regions, the cap may be set high and emissions permits may be allocated freely at the outset of a cap-and-trade program or, in the case of a carbon tax, certain sectors and emissions sources may be excluded for a period of time. The liberal authorization of offset purchases or tax credits for GHG-offsetting activities may dilute the cap or limit the impact of a carbon tax regime. Although such concessions may be necessary in order to gain political consensus for bringing about a pricing program, the result may be less effective pricing coverage and/or a slow phase-in period when opportunities for clean energy investments may be lost.

Impracticality and High Cost of Early and Universal Coverage

Difficulties in measuring emissions reductions, uncertainties about economic impacts, and concerns over risks associated with gaming and cheating are likely to make universal coverage of emissions sources and gases unlikely, at least initially. Additionally, the time delays involved in demonstrating and implementing new technologies that will be needed in some cases to respond efficiently to carbon prices may also create practical impediments to achieving public and political acceptance for early and broad pricing coverage. Incomplete coverage of emissions sources and constrained prices can reduce incentives both to purchase emissions-reducing technologies and to engage in emissions-reducing behavior.

Barriers to a Timely and Robust Response to Pricing

Even if pragmatic considerations can be addressed in ways that produce a comprehensive carbon-pricing program, complementary policies may be desirable to bring about a more timely and robust response to pricing. Specifically, complementary policies may be needed to address the following:

- *Weak R&D investment signals.* Market signals must be strong and sustained to stimulate the large amount of private investment in R&D that will be needed to further the transition to a lower-GHG economy. Private R&D expenditures will provide the foundation for future technologies, but, as discussed in Chapter 5, recent levels of private-sector energy research funding have been low by historic standards. Initial GHG pricing may not be sufficiently high to incentivize greater private investments, particularly in basic research. Complementary policies may be needed to foster greater interest in such research, until the time when pricing signals reach the levels needed to reward investment risk.

- *Long-term capital investments.* Many long-term capital investments (for instance, in the electricity production sector) have lifespans of 50 to 80 years and, thus, can lock in emissions-intensive technologies. Policies to accelerate replacement of existing capital stock (where new technology can yield emissions reductions) would therefore be desirable, but such turnover is less likely to be made if there is uncertainty associated with a carbon-pricing system (i.e., if it may be relaxed or strengthened over time by policy makers).

- *Market failures.* The U.S. Department of Energy identified 20 barriers that have inhibited the deployment of technologies with the potential to reduce GHG emissions (Brown et al., 2007; DOE, 2009). Some of these barriers stem from market failures that limit the response to carbon pricing by consumers and businesses. One example is the *principal-agent asymmetry* barrier (discussed in Chapter 3), which occurs when those subject to the pricing signals are different from those who are capable of responding.

- *Policy failures.* Perverse incentives can arise when other public policies undermine pricing signals. This can occur, for instance, with potentially conflicting tax policies, regulatory inflexibilities that limit technology introduction, and legal restrictions on certain behaviors (such as zoning restrictions that limit high-density development along transit corridors) (Brown and Chandler, 2008). Some of these policies may be necessary to achieve other goals (for instance, vehicle safety standards that mandate reinforced roofs and thus heavier vehicles), but their adverse impacts on GHG emissions-reduction goals can often be modified through strategic complementary policies.

- *Price inelasticities.* Although price inelasticities found in some sectors should not necessarily be defined as a "barrier," they can inhibit or slow the response to pricing policies. For example, studies have concluded that pricing strategies are less effective in reducing GHG emissions from transportation than from electric generation (e.g., see Figure 2.11 in this report). Price inelasticities imply that the good or service in question is considered highly valuable by users, or that the costs of substituting alternate goods or services are very high. Complementary policies may help overcome some of these inflexibilities by lowering substitution costs or making markets more robust for competing technologies.

- *Imperfect or incomplete information.* Individuals (whether consumers, voters, investors, or corporate decision makers) cannot be expected to alter behavior for the purpose of reducing GHG emissions without having an adequate understanding of climate change and its impacts and the response options available. The American public, however, generally suffers from limited information (Coyle, 2005) and inaccurate biases and assumptions about such matters (e.g., Bazerman and Hoffman, 1999; Hoffman and Henn, 2008; Kahneman and Tversky, 1979). As one example, psychologists find persistent belief that there is an inherent trade-off between economic development and climate change limiting strategies, regardless of evidence to the contrary. Complementary policies may be needed to help address these sorts of social and psychological barriers and resistance to change. These policies could include public education programs (akin to antismoking or healthy diet campaigns) and incentives and encouragement for low-GHG-emitting behaviors (e.g., providing incentives for more bike riding through the construction of dedicated street lanes and proper storage facilities).

- *Infrastructure limitations.* Finally, consideration must be given to obstacles to innovation presented by "technological lock-in" of physical infrastructure (Fisher, 2009). Government plays a central role in creating many types of infrastructure and, thereby, has numerous opportunities to enable GHG emissions-reducing technologies. For example, advances in renewable energy and plug-in hybrid vehicles must be enabled by expansion and development of the electricity grid, development of high-speed rail requires the development of new track lines, and the city-scale development of "cool" roofs and pavements and expanded tree cover may require public incentives for private development. In these sorts of examples, government can assist innovation by gaining rights-of-way, altering regulatory structures, or providing broad-scale leadership and direction.

Building a Strategic Portfolio of Complementary Policies

Although many of the carbon-pricing system shortcomings described above could eventually be resolved over time, they need to be considered in light of the urgency and difficulty of meeting a stringent 2050 emissions budget. Delays in beginning emissions reductions, or in creating new technology options for later deployment, will make an extremely challenging task more difficult. For this reason, we believe that a set of carefully targeted complementary policies is justified. However, to avoid compromising the long-term effectiveness of a carbon-pricing policy, complementary policies need to be strategically focused on accelerating near-term emissions reductions and generating long-term technology options. This section presents a short list of goals for complementary policies designed to take advantage of the highest leverage opportunities.

Near-Term, High-Leverage Opportunities

In Chapter 3 we identified the major opportunities for reducing domestic GHG emissions: increasing energy efficiency, accelerating the introduction of renewable energy sources, advancing demonstration of advanced commercial-scale nuclear power, developing and deploying CCS technology, and advancing low-GHG-emitting transportation options. In the sections below, we describe how appropriate complementary policy interventions could help overcome existing barriers in each of these areas. Examples of specific policies are presented in Table 4.1.

- *Increasing energy end-use efficiency.* Exploiting opportunities to enhance efficiency in the use of electricity and fuels offers some of the largest near-term opportunities for GHG reductions (DOE, 2009). These opportunities can be realized at a relatively low marginal cost, thus leading to an overall lowering of the cost of meeting emissions-reduction goals. Furthermore, achieving greater energy efficiency in the near term can help defer new power plant construction while low-GHG technologies are being developed. Although the potential for cost-effective opportunities to reduce electricity and fuel use is large, a variety of market imperfections will reduce the effectiveness of carbon pricing as an incentive for industries and investors to take advantage of these opportunities (Brown et al., 2009b; Gillingham et al., 2009; Tietenberg, 2009). These imperfections include principal-agent problems (discussed in Chapter 3), conflicting incentives created by existing law and regulations, incomplete or imperfect information, and limited access to capital, especially for energy upgrades to industrial facilities—all of which limit the ability of a pricing system to operate as effectively as it otherwise would. In addition, the electricity distribution infrastructure limits the ability of suppliers to transmit real-time price signals to customers.
- *Accelerating deployment of renewable energy sources.* Renewable energy sources, especially wind and solar power, have grown significantly in the past few years. Some renewable energy sources are competitive (or near competitive) with conventional sources even in the absence of carbon pricing, yet it is likely that a continued policy impetus will be required to encourage the widespread adoption of new renewable technology. Capital costs for renewable energy technologies have declined considerably over the past decades,

but they remain a constraint to widespread market penetration. Cumbersome permitting, hookup fees, and interconnection standards are impediments to deployment in many states. In addition, many renewable energy sources are remote from load centers and will require new transmission infrastructure.

- *Determining the cost and performance of new commercial nuclear power.* The *America's Energy Future* (AEF) study concluded that new-generation nuclear power can become a commercially available option by 2020 if large demonstration plants are built without delay (for details, see Chapter 3 and NRC [2009a]). However, significant barriers prevent these demonstrations from taking place without government intervention. Among the more pressing of these are the high costs associated with design and construction of the first few plants and associated financial risk to investors, uncertainty with respect to new licensing procedures, constraints in the supply chain infrastructure, uncertainty regarding long-term waste disposal, and the possible shortage of trained workers. Expansion of nuclear power globally is further complicated by concerns about nuclear proliferation. No new nuclear plant has been built in this country in 30 years, and the cost of a new plant today is so large as to create substantial risk for private investment.

- *Determining the cost and performance of coal with CCS.* The AEF study likewise concluded that coal with CCS could become commercially available by 2020 if we move ahead immediately with construction of full-scale demonstration plants. Technical and other uncertainties surrounding CCS create similar risks for investment. Some of the principal risks are associated with the safety and permanence of geologic storage. Uncertainties also surround the legal frameworks for property rights (to underground pore space), regulatory requirements for site permitting and operation, and long-term liabilities after site closure. A relatively low carbon price, even if it were in place today, would be insufficient to overcome these risks in time to complete an accelerated demonstration program without government assistance.[18]

- *Advancing low-GHG transportation options.* Near-term opportunities exist to reduce GHG emissions from the transportation sector, primarily through increased vehicle efficiency and the use of low-GHG alternative transportation fuels. As household incomes rise, this changes the way people value time and causes demand for motor vehicle travel to become increasingly less responsive (inelastic) to changes in fuel prices (Small and Van Dender, 2007). For instance, a price of $100 per ton of CO_2 translates into about $1 per gal-

---

[18] NRC (2009a) concluded that, even with high carbon prices (over $100/ton) to stimulate private-sector research, government funding could speed the attainment of R&D goals by 3 years.

lon of gasoline. Until CO_2 prices reach higher levels, complementary policies would be needed to realize significant near-term GHG reductions from the transportation sector. Such policies may deal directly with vehicle efficiency (in particular, ensuring that the fuel efficiency standards mandated in the 2007 Energy Independence and Security Act [EISA] are met as quickly as possible), incentivizing the use of efficient modes of passenger and freight traffic as well as investments in new infrastructure, and advancing low-GHG fuels.

Support of Basic Research

Chapter 3 discussed the urgency of conducting research into new technologies. The private sector is likely to under-invest in basic research, as the investing firms cannot fully capture the resulting benefits of such efforts. Cohen and Noll (1991) describe some of the reasons why private firms are unable to realize the benefits of basic research; for example, the benefits gained by "free riders" may dilute the benefits to the firms incurring the cost of research; the investing firm may create an innovation that turns out to be of no strategic relevance to the firm itself, leaving little motivation to pursue it; or the financial risk of pursuing a new technology may be unacceptably high. A carbon-pricing system may help mitigate this problem by creating a benefit for private-sector research into emissions-saving technologies, but a pricing system alone is not likely to result in adequate investment in basic research (Fisher, 2009). Because basic research is the necessary first step in developing many new technologies, direct governmental funding support will remain an essential complementary policy. Chapter 5 discusses the nature of the government role.

Managing Asset Turnover in the Energy Sector

Developing and deploying new technology that limits GHG emissions is essential, but decarbonizing the energy sector also requires retrofitting, retiring, or replacing embedded carbon-intensive infrastructure. The turnover of existing infrastructure can be very slow because existing capitol stock, as well as capital-intensive industrial process infrastructure, can have a long lifetime—ranging, for example, from ~15 years (light-duty vehicles) to ~50 years (refineries, including upgrades). In the absence of policies to encourage turnover, the Energy Information Administration (EIA) anticipates retirements of carbon-emitting electricity generators of less than 5 percent by 2030.[19] Aggressive implementation of energy-efficiency measures could significantly reduce

---

[19] EIA (2009) forecasts that less than 5 percent of the domestic electricity-generating capacity existing in 2007 would be retired by 2030. Almost all of these retirements would be natural gas plants.

overall energy demand, but this would not necessarily stimulate much turnover of existing infrastructure and in fact could even reduce the rate of turnover, since there is less need to add new capacity.

Achieving the 2050 emissions budget means that it will be essential to have clear and credible policies for retiring or decarbonizing much of the current and planned emissions-intensive infrastructure. Ideally, a carbon-pricing system would define this path, but, as discussed earlier, the initial pricing mechanism may be inefficient or slow to take shape, and the initial prices may be too low to provide a sufficiently strong incentive to ensure retrofitting, replacement, or retirement of existing infrastructure.

Various alternative strategies to accelerate equipment and infrastructure turnover have been advanced, but there are few available examples of successful policy intervention. Achieving accelerated turnover requires forcing retirement of equipment and infrastructure that otherwise would continue in productive use, either by law or regulation or by subsidizing early retirement. The subsidies required might be quite large, since existing equipment and infrastructure must cover only its marginal cost to continue to operate profitably, whereas new equipment and infrastructure must offer the promise of covering its total cost.

This challenge is illustrated by the 2009 "cash for clunkers" program, which provided subsidies for scrapping older, less fuel-efficient vehicles and replacing them with newer, more fuel-efficient vehicles. While the program temporarily stimulated the purchase of newer, more fuel-efficient vehicles, its ability to generate a long-term positive impact on the energy efficiency of the automobile fleet has yet to be demonstrated. Similar programs in Europe, which did not require scrapping of the older vehicles, often led to export of the vehicles for use elsewhere.

Performance standards for fossil-fuel power plants may be a useful regulatory tool for driving change in the electricity production sector (Samaras et al., 2009), although such efforts must be carefully designed to avoid the unintentional result of actually slowing down equipment replacement rates. In the consumer products market as well, policies must be formulated carefully—for instance, to avoid the "second refrigerator option" in which new, more energy-efficient purchases are subsidized while the older technology continues to be used, thus adding to overall energy demand. Examples of effective policy options in this regard include tax credits and direct buy-down of replacements.

If complementary policies for influencing asset turnover are not implemented immediately, industry response to the early carbon pricing system should be closely monitored to determine if additional policy action is required to accelerate infrastructure retirement process. If the pricing mechanism is delayed significantly, complementary policies may be essential.

Other Potential Strategies

The high-leverage opportunities discussed above were selected based on a few key criteria: (1) there is a significant near-term opportunity to reduce emissions, (2) there are circumstances that limit the early effectiveness of the pricing system to realize the potential opportunity, and (3) government can play an effective role in mitigating these adverse circumstances. As noted in Chapter 3, there are additional opportunities that will need to be considered to ultimately meet the emissions budget, but many of these are not yet developed to the point of satisfying all of our selection criteria. A few such opportunities appear to be fairly close to the point of making a useful near-term contribution, however. Below we discuss some of these key opportunities on the horizon.

- *Heavy- and medium-duty vehicle fuel efficiency.*[20] The opportunity noted above for improving transportation energy efficiency focuses on light-duty vehicles, because these account for 65 percent of the energy consumption in the transportation sector. Heavy- and medium-duty vehicles are the second largest category of transportation energy consumption, at ~20 percent of the total. At present, no fuel efficiency standards have been set for this vehicle category. The process of setting such standards is under way but too early in its development to know what potential savings can be realized. If the process is successful, it could result in an important near-term contribution to emissions reductions.
- *Agriculture and forestry sequestration of carbon.* The economic potential for sequestering carbon in forests and soils is discussed in Chapter 3. The most likely mechanism for encouraging domestic carbon sequestration in agriculture and forestry is through the use of domestic offsets purchased by primary GHG emission sources, but this may need to be complemented by other policies in order to capture certain opportunities. In some cap-and-trade programs the high transaction costs involved have excluded certain target groups (particularly small emitters) from the market. Programs encouraging sequestration at a local level, managed by agencies such as the U.S. Department of Agriculture Natural Resources Conservation Service and Forest Service, can help reach

---

[20] Medium and heavy trucks are trucks with a manufacturer's gross vehicle weight exceeding 10,000 pounds (medium, 10,001 to 26,000 pounds; heavy, over 26,000 pounds).

the potential for activities such as forest management and small farm agricultural sequestration. Avoiding deforestation can be encouraged by imposing land-clearing offsets, as is done with existing wetlands programs. A number of practical issues need to be addressed for the market to function well (e.g., dealing with transaction costs, leakage, property rights, and additionality concerns). Careful attention to these structural issues would accelerate the contribution of agriculture and forestry CO_2 sequestration to meeting the overall emissions-reduction budget.

Reducing deforestation at the international level is another possibility and is the focus of mechanisms for reducing emissions from deforestation and forest degradation (REDD). As explained by Murray et al. (2009b), making REDD viable on a large scale implies that there is demand for international forest carbon reductions and a sufficient supply capacity of forest carbon credits to meet this demand at a price that is competitive with other mitigation options. In addition, this needs to be coupled with provisions to address the possibility of emissions leakage and impermanence, plus infrastructure (e.g., technological and legal) to ensure that reductions are properly quantified and monitored. Finally, the rights to REDD payments must be properly established. Numerous recent studies provide insights into the potential costs and impacts of a global market for REDD credits (e.g., Boucher, 2008; Busch et al., 2009; Kindermann et al., 2008; Sohngen et al., 2008).

- *Non-CO$_2$ GHGs.* Non-CO_2 GHGs could play a significant near-term role in the overall U.S. emissions-reduction effort; in most cases, including these gases in the overall GHG pricing system is likely to be the most efficient means to encourage action. But as discussed in Chapter 6, there are significant benefits to also pursuing complementary international efforts to reduce emissions of gases such as methane and hydrofluorocarbons.

Examples of Complementary Policy Options

The question of what specific complementary policies are best suited to address the goals discussed above is the subject of considerable study and debate and is ultimately a decision for policy makers. Rather than try to provide a comprehensive list of options and assessments of each, Table 4.1 offers a series of examples that are illustrative of what can be done, in terms of both mandatory regulatory standards ("sticks") and voluntary incentives ("carrots"). This includes policies for advancing technology and policies for influencing individual behavior and consumer choices. For each policy, the table details some pros and cons regarding its effectiveness in addressing the kinds of shortcomings and barriers associated with GHG pricing.

TABLE 4.1 Illustrative Options for Meeting Complementary Policy Goals

Option	Pro	Con
Policies for Increasing the Efficiency of Electric Energy Use		
	Mandatory Regulatory Standards	
Building performance standards for new construction	Building codes can overcome market information and incentive problems to promote new efficient buildings where the energy savings more than cover the up-front cost. Building codes in many states have not been updated for years.	May increase up-front costs for new buildings. Efforts to update building energy codes often result in protracted political conflict. Raises intergovernmental issues if the legislation is instituted at the federal or state level and enforcement is at the local level.
Building performance standards for existing construction	Improvements in existing building stock can yield financial benefits to building owners. Retrofit programs would be consistent with economic stimulus programs already under way.	May produce resistance from building owners over imposed up-front costs and extended payback periods.
Expand and intensify appliance efficiency standards	Proven ability to reduce energy use. Current administrative structure already in place.	Will increase initial cost in many cases. Standards must be revised periodically to take advantage of technological advances. Revisions can be time-consuming and countered by vested interests.
Standards for industrial equipment efficiencies (such as combined heat and power)	The industrial sector is the largest end-use sector, consuming more than 50 percent of delivered energy worldwide.	The relatively high costs for industrial energy-efficiency improvements may create opposition. Many large industry users already have incentives to manage energy costs to remain competitive.

Continued

TABLE 4.1 Continued

Option	Pro	Con
Voluntary Policies and Incentives		
Efficiency tax incentives (and other financial incentives: subsidies, rebates, grants, direct installation/upgrade assistance) for homeowners, businesses, and building owners	Creates a clear market incentive for investments in low-GHG technologies.	May lower tax revenues in a period of already low revenues due to the economic downturn. Less effective when not stable and predictable over time
Support improvements to the electricity grid through smart grid or national grid technologies[1]	Necessary complement for the effective use of dispersed renewable sources. Can yield greater efficiencies in energy use. Can promote more informed consumer electricity choices.	Siting of new high-tension power lines is likely to face local opposition. Developing cost-sharing arrangements to fund grid investments can produce conflict among states.

Policies for Increasing the Energy Efficiency of Transportation

	Mandatory Regulatory Standards	
Higher motor fuel taxes	Will create an additional economic incentive to reduce vehicle miles traveled (VMT) and purchase more fuel-efficient vehicles. Can be a source of needed funding from users for the currently underfunded transportation infrastructure. Reduced oil demand has benefits for national energy security.	Will directly raise costs for consumers. Could adversely impact the poor, since fuel costs are a higher share of their transportation budget. Must overcome considerable political opposition to raising fuel taxes at all levels of government.

TABLE 4.1 Continued

Option	Pro	Con
Energy efficiency performance standards for new vehicles (i.e., the standards promulgated in the 2007 Energy Independence and Security Act [EISA]).	Will stimulate the development and market deployment of additional low-emission vehicles.	Could raise costs for new vehicles. May promote more VMT by lowering the cost of driving. May be resisted by motorists unless accompanied by demand-inducing measures such as higher fuel prices.

Voluntary Policies and Incentives

Option	Pro	Con
Feebates[2] and other financial incentives to spur consumer interest in energy-efficient vehicles	Creates a clear financial incentive for investment in energy-efficient vehicles. Can stimulate markets for fuel-efficient vehicles.	Tax incentives and other government financial incentives will increase government expenditures.
Investments in transportation infrastructure for more efficient operations	More efficient operations of highways and airways can reduce energy use through fewer delays and less circuitous routing.	More efficient operations may increase demand and shift traffic from less energy-intensive modes such as rail.
Promote "smart growth" initiatives	Reduces dependence on automobiles and increases use of public transportation; reduces urban sprawl.	Some developers may oppose efforts that appear to restrict market opportunities. May face barriers from existing zoning codes and lack of regional-scale power over land-use decisions.

Continued

TABLE 4.1 Continued

Option	Pro	Con
Policies for Accelerating Deployment of Renewable Energy Sources		
	Mandatory Regulatory Standards	
Adopt national renewable portfolio standards	Provides a flexible way to accelerate the deployment of renewables. Precedent exists at the state level.	Common federal floor may be much harder to reach in some states than others (depending in part on whether renewable energy credits are tradable among states). May require statutory alteration in federal/state oversight responsibilities for energy management.
Adopt national feed-in tariff legislation	Promotes electricity supplied to the grid from renewable sources by providing a guaranteed price. Considerable experience with their use from Europe. Provides price stability for qualifying energy sources, which leads to increased investment and deployment.	May hit poor the hardest unless correcting policies are enacted (i.e., feebates). Careful structuring of price mechanism is required to avoid unnecessary cost transfers onto consumers.
	Voluntary Policies and Incentives	
Enhance the development and deployment of cellulosic biomass and biofuel	Private investments in biomass development can be increased through government financial support for basic research to prove its scalable deployment.	Suffers from a high perceived risk related to the availability of long-term supply and commercial viability.
Support improvements to the electricity grid through smart grid or national grid technologies	Necessary complement to the effective use of dispersed renewable sources. Can yield greater efficiencies in energy use. Can promote more informed consumer electricity choices.	Siting of new high-tension power lines is likely to face local opposition. Developing cost-sharing arrangements to fund grid investments can produce conflict among states.

TABLE 4.1 Continued

Option	Pro	Con
Stable production tax incentives or other renewables tax supports, including incentives for distributed generation and cogeneration	Creates a stable and predictable financial incentive for long-term investment planning.	May create concerns over the government "picking winners" in the market development of energy sources.

Policies for Deploying New Commercial Nuclear Power and Coal with Carbon Capture and Storage

Voluntary Policies and Incentives

R&D for carbon capture sequestration	CCS offers a technological option for reducing GHG emissions from a critical U.S. energy source. Can reduce the burden that will fall on coal-producing and coal-dependent states. Can produce a national security benefit by allowing greater dependence on domestic energy sources. May pave the way for greater cooperation from coal-reliant countries (e.g., China, India). Can reduce overall policy portfolio costs.	Capture technologies are still not demonstrated in commercial full-scale power plant operations. Is relatively costly.
Help underwrite the risk of constructing initial "new nuclear" power plants	Nuclear is an alternative to more carbon-intensive sources of base load power. Would provide the risk mitigation necessary for allowing initial demonstration plants to be built.	Public opposition to nuclear power remains high. Financing hurdles for the construction of nuclear power remain high and uncertain. Waste disposal and proliferation problems must be resolved.
Develop long-term solutions for nuclear waste disposal	Sound nuclear waste policies are a critical enabling factor in the long-term deployment of nuclear power.	Considerable public and political opposition to a national waste repository and the transportation of nuclear waste over long distances.

Continued

TABLE 4.1 Continued

Option	Pro	Con
Policies for Decarbonizing Transportation Fuels		
	Mandatory Regulatory Standards	
Low-carbon fuel standard	Can complement higher fuel prices by encouraging investment in low-carbon fuel sources. Can incentivize fuel suppliers to reduce carbon at all stages of the fuel production cycle.	May lead to increased use of biofuels from food-based feedstocks. Risk of leakage unless the standard is applied nationally. Challenges in monitoring and accounting for claimed emissions reductions by suppliers at various stages of the fuel production cycle.
	Voluntary Policies and Incentives	
Tax incentives or subsidies for the supply of low-GHG biofuels	Biofuels are a rapidly expanding market with great opportunity for contributing to domestic supply of energy.	If emphasis is not placed on low-GHG sources (e.g., cellulosic feedstocks), it could stimulate increased production of fuels produced from food-based feedstocks or grown on land that could have been used to produce food; it may have little or no net benefit for GHG emissions.
R&D support for the development of low-carbon vehicle propulsion systems	New drive-train technologies such as hydrogen, plug-in hybrids, electrics, and battery storage can figure prominently in both reducing GHG emissions and enhancing future economic competitiveness.	The track record of government picking specific technologies to achieve a goal is poor.
Investment in infrastructure to support the development and use of alternative vehicles and energy sources	Public support for infrastructure to aid in the distribution and delivery of alternative energy sources (e.g., biofuel, hydrogen) can accelerate deployment of the technology.	Can create long-term lock-in if infrastructure investments are misdirected. Risk of government choosing "winners" and discouraging competing technologies with more promise.

TABLE 4.1 Continued

Option	Pro	Con
Policies for Managing Asset Turnover		
	Mandatory Regulatory Standards	
Performance standards for new coal-fired power plants	Can prevent new coal plants from using a disproportionate share of the emissions budget. Can create markets for CCS technologies and stimulate R&D and technology innovation (see Chapter 5).	May increase cost for new coal power plants. May create incentive to keep old plants running.
Performance standards for existing coal-fired power plants	Can prevent existing plants from using a disproportionate share of the emissions budget. Can create markets for CCS retrofit technologies and stimulate technology innovation. Can accelerate the retirement of older plants that are not amenable to emission controls.	Can increase the cost to consumer for energy, yielding public and political opposition. Political opposition in regions with many coal-related jobs.
	Voluntary Policies and Incentives	
Incentives for the retirement of inefficient vehicles	Will stimulate the market for fuel-efficient vehicles at a time when the automobile sector is in financial distress.	Political opposition may view this as a subsidy for the auto sector. Could affect the poor by removing lower-cost vehicles from the market. Recent experience with "cash for clunkers" shows this may be an inefficient approach to influencing demand for fuel efficient vehicles at reasonable cost (e.g., Knittel, 2009).

Continued

TABLE 4.1 Continued

Option	Pro	Con
Policies to promote urban redevelopment	Will stimulate the market for urban real estate, revitalizing communities. Will promote the use of public transportation and complement "smart growth" initiatives by reducing demand for sprawl.	May require considerable funding in communities where market signals do not promote development activity on their own. Development benefits may be slow to emerge.

[1] Smart grid technology supports both energy efficiency and the deployment of renewable technology, and so appears in both sections of Table 4.1.

[2] A policy where a fee is levied on the purchase of "gas guzzling" vehicles (e.g., via registration fees, surcharge on initial vehicle purchase) and the money is put toward rebates for purchasers of highly efficient vehicles.

Whatever options are selected, policy makers will have an ongoing task of evaluating their effectiveness and efficiency, and adjusting them to changing circumstances, especially in order to avoid unnecessary conflict with an evolving carbon-pricing mechanism. For instance, some policies adopted to complement the pricing mechanism in its early years may eventually outlive their usefulness, because over time the incentive created by a pricing mechanism should become sufficiently strong to elicit the most efficient response from private markets. When that happens, the complementary policies are no longer required. Thus, the design of complementary policies should include consideration of their eventual phase-out.

On the other hand, if the pricing mechanism fails to evolve to the point of providing the appropriate incentive, then the complementary policies may have to be continued, or new policies may need to be phased in (for example, in the case of managing the retirement of existing electric power plants). In the extreme, if the pricing mechanism is not enacted or is abandoned, then the complementary policies would become the foundation of our nation's strategy to meet the 2050 GHG budget; thus, the policy portfolio would have to be adjusted accordingly.

As discussed in Chapter 3, it is useful to look across the traditional sectors of GHG emissions-reduction efforts (such as those shown in the preceding tables) and consider the perspective of who is responsible for relevant decisions and actions. Chapter 3 discussed the opportunities that exist for influencing individual or household-level

---

BOX 4.2
Policy Strategies for Reducing Household-Level GHG Emissions

As discussed in Chapter 3, GHG emissions from U.S. households could be far lower with changes in how people adopt and maintain energy-using equipment both inside the home (i.e., appliances) and outside the home (i.e., cars) (Bressand et al., 2007; Dietz et al., 2009; Hirst and O'Hara 1986). The main policy options for encouraging these sorts of changes and reducing household emissions are discussed below.

Regulations in the form of efficiency standards for homes, appliances, and automobiles have in some cases successfully changed the product mix and increased overall efficiency, although they have not altered the trend toward larger units with more energy-using features. Standards are an effective option for new equipment, but they generally do not force upgrades or retrofits of existing equipment; in some cases, standards can even strengthen incentives to prolong the life of old, inefficient equipment. This need not always be the case, however. For instance, a study by the California Energy Commission found positive benefits of regulations requiring building energy upgrades at the time of sale (California Energy Commission, 2005), and several localities (e.g., Berkeley and Austin, California) have begun to adopt codes requiring some combination of home energy rating and retrofit (City of City of Austin, 2009; Berkeley, 2008.).

Economic influences have highly variable effects on consumers. This can be seen in the tremendous variations in implicit discount rates for energy efficiency that have been calculated from studies of appliance purchases (Ruderman et al., 1987) and in the large variation in the proportions of homes that are found to make energy-efficiency improvements in response to financial incentives (Stern et al., 1986). With appliances, much of the variation is due to the fact that it is often not the consumer who makes the actual choice, but a builder or repair professional. With home retrofit incentives, it seems to be due to attributes of the organization administering the program and of its implementation (Gardner

---

choices and behavior related to energy use. Box 4.2 considers the types of policy interventions that can help ensure those opportunities are effectively pursued.

INTEGRATING THE POLICY OPTIONS

The nation needs a strategic, integrated strategy for evaluating and selecting the most effective portfolio of policy options. In Chapter 1 we suggested a range of principles or criteria that could be used to evaluate all policies on an individual basis. The first four of those criteria may be particularly important in the policy-making arena; this includes the criteria of policies that are environmentally effective, are cost-effective, help stimulate innovation, and promote equity and fairness of outcome. In addition, below we suggest a set of "ensemble" criteria as guidance for finding a balanced, effective portfolio of policies:

and Stern, 2002; Stern et al., 1986).

Communication instruments generally have had very limited effects on energy use and emissions (Abrahamse et al., 2005; Gardner and Stern, 2002; NRC, 2002a). Generic information, such as is offered in many mass media energy campaigns, has had little effect on behavior or energy consumption. Interventions such as eliciting a personal commitment or using neighbors as behavioral models can be quite effective but are not readily transferable into widespread policy. Targeted information, such as daily feedback on household energy use, has produced savings in the range of 10 percent of household use of a target fuel (usually electricity). These savings usually result from adjustments in the use of household equipment (e.g., lower temperature settings on hot water or shorter showers) rather than changes in equipment stocks.

The most effective policy interventions combine multiple approaches in order to address multiple barriers to behavioral change. For example, 85 percent of the homes in Hood River, Oregon, underwent major energy efficiency retrofits in 27 months under a program that provided large financial incentives, convenience features (e.g., one-stop shopping), quality assurance (e.g., certification for contractors, inspection of work), and strong social marketing (Hirst, 1988). Similarly structured programs have produced penetrations of up to 19 percent per year in other communities, although the same incentives with different implementation have yielded penetrations under 2 percent per year (Stern et al., 1986).

A key lesson learned from these experiences is that policy instruments are most likely to be effective when they "provide just what is needed to overcome the barriers to obtaining the [policy] objective" (Stern, 2002). For a sector facing multiple barriers, multipronged interventions can be far more effective than financial incentives or information alone. Well-designed policy interventions aimed at households can likely also increase the speed of adoption of new emissions-reducing household technology and promote household choices that contribute indirectly to reducing GHG emissions.

- *Widespread participation.* The portfolio of policies should be designed to draw in vigorous action at all levels, from household and individual, to state and local, to international.
- *Temporal effectiveness.* The mix of policies should stimulate immediate action and payoffs but also be consistent with long-term goals. Short-term priorities may focus on stimulating behavior change, deployment of available technologies, and capital stock turnover. Such efforts need to be complemented with longer-term priorities such as greater support for basic R&D and associated innovation policies.
- *Comprehensiveness.* The mix of policies should lead to comprehensive coverage of GHGs, strategies, and major sectors of the economy.

The major strategic elements of policy integration involve recognizing and capitalizing on the interactions among policies, sequencing policies for maximum cost-effectiveness, and taking synergies among different policy goals into account. Each of these is discussed below.

Policy Portfolio Interactions and Sequencing

It is important to consider the interactions among different emissions-reduction policy instruments, both to identify and capitalize on opportunities where the joint outcomes are greater than the sum of the independent parts and to anticipate circumstances where the joint effect may diminish the emissions-reduction effort or even be counterproductive. Below is one example that illustrates the complexity of these types of interactions.

Renewable Portfolio Standards (RPSs) and Renewable Fuels Standards or Low Carbon Fuel Standards are examples of measures that overlap with a GHG pricing policy, in the sense that they all are intended to reduce GHG emissions. If complementary policies were truly redundant with a cap-and-trade system (meaning the emissions reductions would take place in response to carbon prices even in the absence of these complementary policies), then their addition to the policy portfolio would not affect carbon prices or overall program costs. In practice, however, complementary policies would likely force technological choices that would not otherwise occur under a pricing policy alone. This may increase overall program costs, but at the same time may lower the carbon price.[21] Total emissions may not be further reduced by the introduction of complementary policies (since that total is set by the cap), but the source of those emissions is affected. For instance, a recent MIT modeling study (Morris, 2009) found that adding an RPS to a cap-and-trade system forces a higher proportion of electricity generation to come from renewables. This study also found that (as proposed above) an RPS combined with a cap-and-trade policy leads to the same total emissions as cap and trade alone, but at a greater cost despite a lower carbon price (noting that such results can depend on a model's assumptions regarding technological change and other factors).

The MIT study does not envision a large role for offsets, but adding significant offsets to the policy mix would exacerbate the cost impact that the study suggests. By itself, a generous use of offsets would lower emission allowance prices and delay the transition to using renewable energy resources. But this delay could be reduced or eliminated if an RPS is also in place to mandate that a larger proportion of electricity come

---

[21] In a cap and trade, because renewables emit less carbon than the sources they replace, their use lowers the demand for allowances and, hence, lowers the carbon price. Overall program cost is not lowered by this additional use of renewables, however, if they supply energy at a higher cost.

from renewables. This faster transition, however, would increase near-term program costs more significantly than in the scenario without offsets, because the renewables would be more expensive than the offsets they replaced.

Similar considerations affect how policies such as building codes and appliance standards interact with a pricing strategy. In theory, given an appropriate carbon price, all households and businesses would make cost-effective choices, realizing that higher costs of efficient buildings or appliances will be offset by lower expenditures on energy. In practice, however, historical experience shows that information deficiencies and perverse incentives create many circumstances where households and businesses make choices that are not cost-effective. Building codes and appliance standards, if appropriately designed, can help ensure that households and businesses end up making cost-effective choices. However, if the standards are set too low, they might prove to be redundant with carbon-pricing incentives; if the standards are set too high, they might increase program cost by eliminating some cost-effective choices.

A recent analysis of some existing cap-and-trade programs (Hanemann, 2009) argues that the technology innovation stimulated by cap and trade alone will likely be insufficient without also having complementary policies in place. For instance, in order to produce the desired rate of innovation in key sectors, it may be necessary to complement cap and trade with performance standards specifically targeted at those key sectors. This study also suggests that the inclusion of complementary policies can help reduce the possibility that emissions allowance prices will reach a level that is too high to be politically sustainable.

The sequence in which some policies are enacted can affect their outcome. For instance:

- The most important early emissions reductions generally come from energy-efficiency improvements; however, due to the information and incentive gaps noted above, significant improvements may not occur unless policies to complement a carbon price (e.g., building codes, appliance standards, and fuel-economy standards) are put in place early in the process.
- Most scenarios envision an ongoing significant role for coal, but this will not be consistent with the proposed emissions budget goals unless CCS quickly becomes available and policies are enacted to ensure its use.
- The California Air Resources Board anticipates that low carbon fuel standards will work best if developed in concert with technology-forcing regulations designed to reduce GHG emissions from cars and trucks, as well as land-use and urban growth policies designed to reduce transportation-sector emissions (CARB, 2008).

An additional consideration is the interaction of climate change limiting goals with policy goals in other related areas—for instance, adapting to climate change impacts, protecting public health through air pollution mitigation, reducing dependence on foreign oil and advancing energy security, expanding economic development and employment opportunities, and enhancing national competitiveness and international markets for domestic goods. See Chapter 6 for further discussion of these issues.

Because of the complexity of these interactions, there is a diversity of views among experts about the appropriate role of complementary policies. This is another rationale for why it will be necessary to learn from experience, and adapt as needed, as we proceed with implementing a policy portfolio.

Emissions Leakage

Emissions leakage can undermine the efficacy of GHG emissions-reduction efforts in a variety of ways. For instance:

- Leakage can occur when a regulatory scheme covers only a single region or country (or group thereof) and resulting price differentials push the emissions-producing activities into other regions that are not constrained by the same regulatory controls.
- Leakage can occur when efforts to reduce emissions in one sector or location cause a resulting unsatisfied demand that is then satisfied somewhere else, with a consequent rise in net emissions.
- Leakage can be caused by an inappropriate certification for offsets, causing emissions reductions from a project-based offset to be less in practice than claimed.

These are, of course, not really new issues. Basic international trade economic theory has long shown that actions that increase the cost of goods in one country for a traded commodity will cause a countervailing reaction in another country, replacing the production with a shift in market share. Leakage is just an acknowledgment that, when GHG-limiting strategies differentially raise costs and prices, there will be associated production changes, and, in turn, GHG emission patterns change. Namely, climate change limiting policies that displace production in controlled regions will inevitably stimulate additional economic activity (and consequent leakage) in uncontrolled regions. While leakage can, in theory, affect almost any GHG emission source (including agricultural sources; see Box 4.3), some studies indicate that these leakage threats are in fact likely to be quite small overall, and largely manifested in a narrow subset of energy-intensive industries(Pew Center, 2009b). A recent Organisation for Economic

---

BOX 4.3
Leakage in an Agricultural Setting

Much of the public discussion about emissions leakage concerns has focused on energy-intensive industries. However, both domestic and international leakage issues also arise frequently in the context of actions in the agricultural sector. One recent example concerns the land-use impacts of biofuels. Some evidence suggests that high commodity prices caused by diversion of U.S. corn into ethanol production have stimulated foreign competitors with undeveloped land resources to respond by increasing their production (Fargione et al., 2008; Searchinger et al., 2008). This argument suggests that land conversion leads to loss of grasslands, forests, and other valuable ecosystems, causing both current carbon releases and lost future potential for carbon sequestration. Such arguments underlie the controversial indirect land-use adjustments in the EPA's Renewable Fuels Standard analysis.[1]

Other agricultural programs have faced this issue as well. For example, Wu (2000) shows evidence of leakage within the United States in association with land conversion from pasture (in the Conservation Reserve Program). Furthermore, Wear and Murray (2004) and Murray et al. (2004) show that reduced Pacific Northwest deforestation (designed to protect the spotted owl) was matched by accelerated rates of harvest on regional private lands, in the southern United States and in Canada, with total leakage estimates in the neighborhood of 85 percent.

---

[1] Some in the renewable fuels industry charged that EPA overstated the impact of corn ethanol on U.S. food production and thus exaggerated the expansion of new crop planting in forests and savannahs of places such as Brazil. See discussion, for example, in *The Washington Post*, May 6, 2009 (EPA Proposed Changes to Biofuel Regulations).

Co-operation and Development (OECD) study (Nakano et al., 2009) concluded that the emissions embedded in internationally traded goods are only a small percentage of OECD emissions and, hence, the extent of leakage is likely to be very small.

Conceptually, several unilateral border adjustment policy options are available for dealing with emissions leakage stemming from domestic emissions controls, including, for instance,

- Import taxes on products—or equivalently, requiring allowances from imports—with embodied carbon (that is, high levels of GHG emissions generated during their production) can level the playing field for domestic consumption; however, this does nothing to reduce the competitive disadvantage faced by exporters.
- Export rebates return the value of the emissions embodied in exports to exporters so that they do not face a competitive disadvantage in foreign

markets; however, this does nothing to reduce the competitive disadvantage domestic producers face from imports.

- Full border adjustment policies combine these two measures such that, in effect, only the emissions from domestic energy consumption are taxed.

Attempts to rank the desirability of these various approaches have proved inconclusive, since the ranking depends on many context-specific parameters. Simulations do confirm, however, that the largest share of leakage arises from the effects of climate policies on energy prices, and that adjustment policies can mitigate leakage on the margin but are quite limited in their capacity to affect total global emissions reductions (Fischer and Fox, 2009).

Concerns have been expressed about the protectionist implications of these types of measures (Grimmett and Parker, 2008). However, a recent joint report from the United Nations Environment Programme and the World Trade Organization (UNEP and WTO, 2009) suggests that border adjustments could be legal under WTO rules if they were necessary to limit the magnitude of climate change and were applied in a nondiscriminatory way. This issue is discussed further in Chapter 7.

Leakage concerns arise in the context of both domestic and international offsets. In the international context, the Kyoto Protocol GHG accounting system for participating countries is considered only on a national basis (i.e., no consideration of leakage among countries), but leakage is discussed in the context of project-based emissions reductions such as the CDM. Murray et al. (2005) argue for the importance of developing methods to design projects to minimize leakage, to monitor leakage after projects are implemented, to quantify the magnitude of leakage when it exists, and to take leakage into consideration when estimating an activity's net GHG reduction benefits.

To alleviate leakage concerns for offsets, emissions-reduction projects thus need to be evaluated under broad national and international accounting schemes that consider both direct and indirect implications of project implementation. That is, project evaluations should look not just at the project itself but also at the related impacts in major competitive regions. Some specific strategies for addressing leakage-related offset projects that have been proposed include the following:

- Reduction of the quantity of offsets that can be credited and sold, to account for external leakage, and use of a "leakage discount factor" in the price paid for emissions allowances (see Murray et al., 2004);
- Use of GHG offsets that avoid displacing marketed goods by using less competitive items; for instance, in the context of renewable fuels, focusing on the

use of marginal lands and on the use of municipal, agricultural, and forestry wastes or residues;

- Associated complementary policies, such as avoided deforestation, that independently address leakage; and
- Attempts to extend GHG price signals to more comprehensive global coverage, such that all relevant parties face the same signals and leakage becomes a liability in areas where it occurs.

KEY CONCLUSIONS AND RECOMMENDATIONS

Evidence suggests that a carbon-pricing strategy is a critical foundation of the policy portfolio for limiting future climate change. It creates incentives for cost-effective reduction of GHGs and provides the basis for innovation and a sustainable market for renewable energy resources. An economy-wide pricing policy would provide the most cost-effective reduction opportunities and lower the likelihood of significant emissions leakage, and it could be designed with a capacity to adapt in response to new knowledge.

Options for a pricing system include taxation, cap and trade, or some combination of the two. Both systems face similar design challenges. On the question of how to allocate the financial burden, research strongly suggests that economic efficiency is best served by avoiding free allowances (in cap and trade) or tax exemptions. On the question of how to use the revenues created by tax receipts or allowance sales (or the value of the allowances themselves), revenue recycling could play a number of important roles, for instance, by supporting complementary efforts such as R&D and energy-efficiency programs, by funding domestic or international climate change adaptation efforts, or by reducing the financial burden of a carbon-pricing system on low-income groups.

In concept, both tax and cap-and-trade mechanisms offer unique advantages and could provide effective incentives for emissions reductions. In the United States and other countries, however, cap and trade has received the greatest attention, and we see no strong reason to argue that this approach should be abandoned in favor of a taxation system. In addition, the cap-and-trade system has features that are particularly compatible with others of our recommendations. For instance, it is easily compatible with the concept of an emissions budget, and more transparent with regard to monitoring progress toward budget goals. It is also likely to be more durable over time, since those receiving emissions allowances have a valued asset that they will likely seek to retain.

High-quality GHG offsets can play a useful role in lowering the overall costs of achieving a specific emissions reduction by expanding the scope of a pricing program and offering a financing mechanism for emissions reductions in developing countries. Those gains, however, would be valid only for cases where offsets are real, additional, quantifiable, verifiable, transparent, and enforceable.

Pricing GHGs is a crucial but insufficient component of our nation's climate change response strategy. Because a national carbon-pricing system takes time to develop and mature into an effective market stimulus, and because perverse incentives and information deficiencies can limit the effectiveness of a carbon-pricing policy in practice, a strategic combination of well-targeted complementary policies will be needed. Complementary policies should be focused on advancing the following major objectives:

1. Realize the practical potential for near-term emissions reductions. End-use energy demand and the technologies used for electricity generation and transportation together drive the majority of U.S. CO_2 emissions. Key near-term opportunities for emissions reductions in these areas include the following:

- *Increase energy efficiency.* Enhancing energy-use efficiency offers some of the largest near-term opportunities for GHG reductions. These opportunities can be realized at a relatively low marginal cost, thus leading to an overall lowering of the cost of meeting the 2050 emissions budget. Furthermore, achieving greater energy efficiency in the near term can help defer new power plant construction while low-GHG technologies are being developed.
- *Increase the use of low-GHG-emitting electricity generation options, including the following:*
  - *Accelerate the use of renewable energy sources.* Renewable energy sources offer both near-term opportunities for GHG emissions reduction and potential long-term opportunities to meet global energy demand. Some renewable technologies are at and others are approaching economic parity with conventional power sources (even without a carbon-pricing system in place); however, continued policy impetus is needed to encourage their development and adoption. This includes, for instance, advancing the development of needed transmission infrastructure, offering long-term stability in financial incentives, and encouraging the mobilization of private capital support for RD&D.
  - *Address and resolve key barriers to the full-scale testing and commercial-scale demonstration of new-generation nuclear power.* Improvements in nuclear technology are commercially available, but power plants using this technology have not yet been built in the United States. Although such plants

have a large potential to reduce GHG emissions, the risks of nuclear power are also of significant concern and need to be successfully resolved.

- ○ *Develop and demonstrate power plants equipped with carbon capture and sequestration technology.* Carbon capture and sequestration could be a critically important option for our future energy system. It needs to be commercially demonstrated in a variety of full-scale power plant applications to better understand the costs involved and the technological, social, and regulatory barriers that may arise.

- *Advance low-GHG-emitting transportation options.* Near-term opportunities exist to reduce GHGs from the transportation sector through increasing vehicle efficiency, supporting shifts to energy-efficient modes of passenger and freight transport, and advancing low-GHG fuels.

2. Accelerate the retirement, retrofitting, or replacement of emissions-intensive infrastructure. Transitioning to a low-carbon energy system requires clear and credible policies that enable not only the deployment of new technologies but also the retrofitting, retiring, or replacement of existing emissions-intensive infrastructure. If immediate action is not initiated, the existing emissions-intensive capital stock will rapidly consume the U.S. emissions budget.

3. Create new technology choices. See discussion in Chapter 5.

CHAPTER FIVE

Fostering Technological Innovations

Any successful strategy to significantly reduce greenhouse gas (GHG) emissions will require actions not only to deploy low-emission technologies that are available now but also to foster innovations on new technologies, many of which have not yet been invented, commercially developed, or adopted at a significant commercial scale. Indeed, a study by the Pew Center on Global Climate Change (Alic et al., 2003) concluded that "large-scale reductions in the GHGs that contribute to global climate change can only be achieved through widespread development and adoption of new technologies." Accordingly, much interest in recent years has focused on ways to foster innovation and, in particular, on the role that governments can and should play in that process.

Although research and development (R&D) is a major element of the innovation process, there is growing recognition that technological change requires more than just R&D. Rather, "technological innovation is a complex process involving invention, development, adoption, learning and diffusion of technology into the marketplace. Gains from new technologies are realized only with widespread adoptions, a process that takes considerable time and typically depends on a lengthy sequence of incremental improvements that enhance performance and reduce costs" (Alic et al., 2003).

What strategies and policies can most effectively foster technological innovations that help reduce future GHG emissions, both domestically and globally? To answer this question, this chapter explores the ways in which technological innovations can affect future GHG emissions, the nature of the technology innovation process and the factors that influence it, and the roles of government and the private sector in bringing about desired innovations and changes in technological systems.

THE ROLE OF TECHNOLOGICAL INNOVATION

As discussed in Chapters 2 and 3, GHG emissions depend strongly on the types of energy sources and technologies that are used to provide the goods and services that society seeks. Technological innovations can thus affect GHG emissions in many different ways. For example, new or improved technologies can

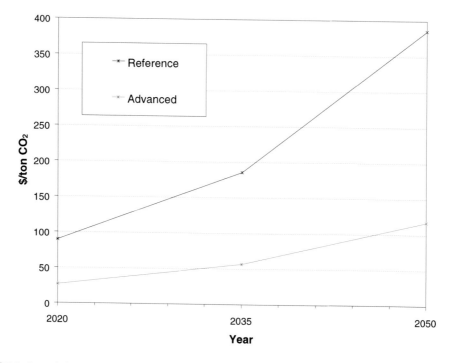

FIGURE 5.1 A model projection of the future price of CO_2 emissions under two scenarios: a "reference" case that assumes continuation of historical rates of technological improvements, and an "advanced" case with more rapid technological change. The absolute costs are highly uncertain, but studies clearly indicate that costs are reduced dramatically when advanced technologies are available. SOURCE: Adapted from Kyle et al. (2009).

- enable a device—whether a vehicle, machine, or appliance—to use energy more efficiently, thereby reducing its energy use and GHG emissions per unit of useful product or service (such as a vehicle mile of travel).
- create or utilize alternative energy carriers and chemicals that emit fewer GHGs per unit of useful product or service (e.g., renewable energy or new fertilizers).
- create alternative means of meeting needs, in ways that are less GHG-intensive, for instance, by using substitute products or materials, by changing agricultural practices, or by making broader systems-level changes such as replacing vehicle and air travel with teleconferencing, or using Internet-based delivery services in lieu of traveling to a store.

Efforts to stimulate technological innovation must be broad enough to affect this full range of possibilities and may also encompass innovations in social and institutional systems that help reduce energy demand and GHG emissions (for instance, through innovations in urban planning and development). Figure 5.1 shows one estimate of how technological innovations can reduce the future cost of GHG emissions reduction. In this modeling study, a "business as usual" case—which assumes a continuation of historical rates of technological improvements—is compared to a case with more rapid technological change. The cost of meeting a stringent emissions-reduction requirement is reduced dramatically when "advanced technologies" are available.

THE PROCESS OF TECHNOLOGICAL CHANGE

Technological innovation is a component of the broader process of technological change, which involves a number of stages (Figure 5.2). Different terms are used in the literature to describe these stages, but four commonly used descriptors are

Invention: Discovery; creation of knowledge; new prototypes.

Innovation: Creation of a commercial product or process.

Adoption: Deployment and initial use of the new technology.

Diffusion: Increasing adoption and use of the technology.

The first stage is driven by R&D, including both basic and applied research. The second stage—innovation—is the term often used colloquially to describe the overall process of technological change. But, as used here, innovation refers only to the creation of a commercially offered product or process; it does not mean the product will be adopted or become widely used. That requires the product to pass successfully through the last two stages—adoption and diffusion. Those two stages are inevitably the most critical to reducing GHG emissions. Large-scale change also must be considered from a "systems" perspective, because the success of any new technology is often dependent upon many other technological and nontechnological factors.

Rather than being a simple linear process, the different stages of technological change are highly interactive, as depicted in Figure 5.2. Technological innovation is stimulated not only by support for R&D but also by the needs and opportunities that emerge from the experience of early adopters, and from the knowledge and experience gained as a technology diffuses into the marketplace. Thus, "learning by doing" is often critical to the adoption and diffusion of new technologies by helping to improve their performance and/or reduce their cost (a process commonly characterized as a learning curve).

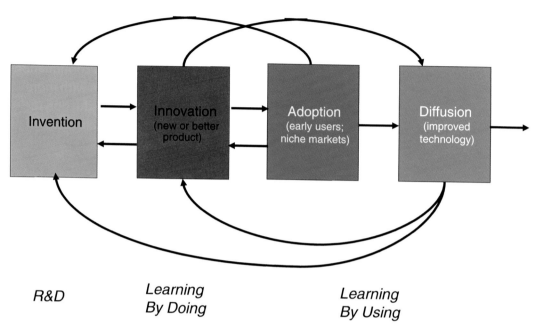

FIGURE 5.2 Stages of technological change and their interactions. The processes of adoption and diffusion typically involve a continuing series of inventions and innovations that require new research and development. SOURCE: Rubin (2005).

Each stage of the process requires different types of incentives to promote the overall goal of technological change. An incentive that works well at one stage of the process may be ineffective—or even counterproductive—at another. In particular, the widespread adoption and diffusion of a new technology may require addressing social and institutional issues that affect the nature and pace of technological change (see earlier discussion of this topic in Chapter 3).

The Essential Role of Markets

In the U.S. economy, most production is by private firms and most output is sold to purchasers in the private sector. The existence of a market is thus critical to the adoption and diffusion of a new technology—and thus to the process of technological innovation. This is true even in cases where the government is the primary customer—for instance, in the procurement of military technology. Indeed, government procurement has been a critical tool for enabling new technologies to enter the market (jet aircraft and electronic computers are two prominent examples). Some technological innovations create new markets or expand existing ones, as exemplified by cell phones and other electronic devices. While R&D may play a critical role in the development of such innovations, R&D alone is not sufficient; there must also be a market for the technology.

A major challenge in reducing GHG emissions is that few if any markets exist for many of the needed low-emissions technologies. What utility company, for example, would want to spend money on carbon capture and storage if there is no requirement or incentive to reduce emissions? How many individuals would willingly buy a more fuel-efficient automobile costing far more than a conventional vehicle in order to reduce their carbon footprint? Costly actions by firms or individuals to reduce GHG emissions provide little or no tangible value to that firm or person. Only government action that requires or makes it financially worthwhile to reduce GHG emissions can create sizeable markets for the products and services that enable such emissions reductions. Government actions to create or enhance markets for GHG emissions-reducing technologies are thus a critical element of the technological innovation process.

The Influence of Government Policies

Different policy measures influence technological innovation in different ways. Table 5.1 shows a set of commonly employed technology policy options (including several that were discussed in Chapter 4) that help create markets by providing voluntary incentives for technology development, deployment, and diffusion. It also lists policy options to impose mandatory regulatory requirements, which may be economy-wide or targeted to certain sectors. Studies have documented the ability of regulatory policies to stimulate innovations that reduce GHG emissions; for instance, energy-efficiency standards for appliances such as refrigerators (Rosenfeld and Akbari, 2008), emissions and fuel economy standards for automobiles (Lee et al., 2010), and new source performance standards for power plants (Taylor et al., 2005). These policies create or expand markets for lower-emission technologies by imposing requirements on manufacturers and industrial operations. The policy options outlined in Table 5.1 are revisited later in the chapter in considering current needs for fostering technological innovation.

Major technological changes in the U.S. energy system and other sectors will be needed to reduce GHG emissions significantly, and this will require an infusion of financial and human resources to support each phase of the process depicted in Figure 5.2. Resources that are critical for technology innovation include money for R&D and people with the requisite training, skills, and creativity to innovate. Below we review current U.S. resources in these areas and estimate the magnitude of new

TABLE 5.1 Policy Options That Can Influence Technology Innovation

"Technology Policy" Options[a]			Regulatory Policy Options
Direct Government Funding of Knowledge Generation	Direct or Indirect Support for Commercialization and Production	Knowledge Diffusion and Learning	Economy-wide Measures and Sector or Technology-Specific Regulations and Standards
• R&D contract with private firms (fully funded or cost shared) • R&D contracts and grants with nonprofits • Intramural R&D in government laboratories • R&D contracts with consortia or collaborations	• R&D tax credits • Patents • Production subsidies or tax credits for firms bringing new technologies to market • Tax credits, rebates, or payments for purchasers and users of new technologies • Government procurement of new or advanced technologies • Demonstration projects • Loan guarantees • Monetary prizes	• Education and training • Codification and diffusion of technical knowledge (e.g, via interpretation and validation of R&D results; screening; support for databases) • Technical standards • Technology/industry extension programs • Publicity, persuasion, and consumer information	• Emissions tax • Cap-and-trade program • Performance standards (for emission rates, efficiency, or other measures of performance) • Fuels tax • Portfolio standards

[a] Based on CSPO and CATF (2009).

financial and "human capital" resources needed to support a major initiative on GHG-related technological innovation. In addition, accelerating technological innovations that reduce GHG emissions will require a variety of policy drivers—to promote R&D, to help commercialize and bring new technologies to the marketplace, and to establish and expand markets for low-GHG technologies. Thus, we also review below current U.S. policies available to support these objectives. Based on the findings from this review, we suggest additional policy measures that are most needed.

RESOURCES CURRENTLY AVAILABLE FOR TECHNOLOGY INNOVATION

This section presents current and historical data on funding for R&D in the United States, especially energy-related R&D, and examines trends in workforce training and employment data related to R&D. Where appropriate, comparisons are drawn with other (nonenergy) industries, and with other industrialized countries, to provide additional perspectives on resources currently devoted to or needed to support technological innovations that reduce GHG emissions.

Federal Funding for Energy-Related R&D

According to the National Science Foundation (NSF), federal spending for nondefense R&D in fiscal year (FY) 2008 totaled $55.9 billion (equivalent to $46.2 billion in FY 2000 dollars, the base year used by NSF to report historical trends) (NSF, 2008). Of this, energy R&D accounted for $1.2 billion (FY 2000 dollars), or 2.6 percent of the total. This is a sharp decline from FY 1980, when the budget authority for energy R&D totaled $6.8 billion—approximately 25 percent of all federal (nondefense) R&D spending that year (Figure 5.3). Contrast this to the trend in federal R&D devoted to health over the same period of time (FY 1980 to FY 2008), which grew from $7.0 to $24.2 billion (from about 25 percent to 52 percent of all federal nondefense R&D spending).

The decline in federal energy R&D spending—both in absolute terms and as a percentage of all federal nondefense R&D—reflects the decline in energy as a national priority after the 1970s. In large part this reflects the sharp drop in world oil prices and the increased availability of natural gas supplies in the 1980s and 1990s. During this period, the U.S. economy also underwent structural shifts away from energy-intensive heavy industries toward light industries and service sectors that reduced the national energy needs for economic growth. However, the reemergence of energy as a national priority in recent years has not yet been reflected in a rebalancing of federal R&D spending. By way of illustration, it would require a 20-fold increase in the FY 2008 level of federal energy R&D spending to equal federal health R&D spending, a sixfold increase to equal federal space R&D spending, and a fivefold increase to equal federal "general science" R&D spending.

Recently there has been some increase in energy-related funding. Figure 5.4 shows Department of Energy (DOE) budget authority for R&D (excluding support for basic energy sciences) from FY 1980 through the FY 2010 budget request (Gallagher and Anadon, 2009). The column labeled "ARRA" shows the funding appropriated in the 2009 American Recovery and Reinvestment Act, which (unlike the annual budget fig-

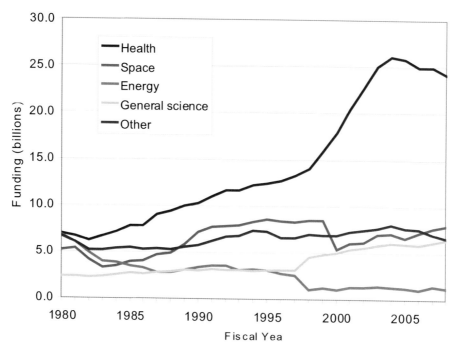

FIGURE 5.3 Federal R&D budget authority by budget function: 1980-2008 (in billions of year 2000 dollars). Over the past decade or so, expenditures for energy R&D have dropped, and they are much lower than for other key areas of science and technology. SOURCE: NSF (2008).

ures in the other columns) appropriated funds to be spent over the next 2 to 5 years. Indeed, the FY 2010 budget request (last column) was devised to complement funds provided by ARRA. Note too that the ARRA provides funding for a new programmatic initiative, ARPA-E (Advanced Research Projects–Energy), as well as $3.4 billion for various clean coal projects.

Figure 5.4 shows that, on an annual basis, total federal energy R&D spending has begun to grow in the past few years, but it is still well below (roughly half) its 1980 level. The figure also illustrates the shifting of priorities within the federal energy R&D budget over time. For example, funding for research on nuclear fission fell from about $1.5 billion (FY 2000 dollars) in 1980 to nearly zero in 1998, before beginning to rise again in 2002. Other categories of spending also experienced wide fluctuations during this 30-year period.

It is also instructive to benchmark U.S. government spending on energy R&D against that of other industrialized countries. To account for the different sizes of national economies, Figure 5.5 (developed from data published by the International Energy

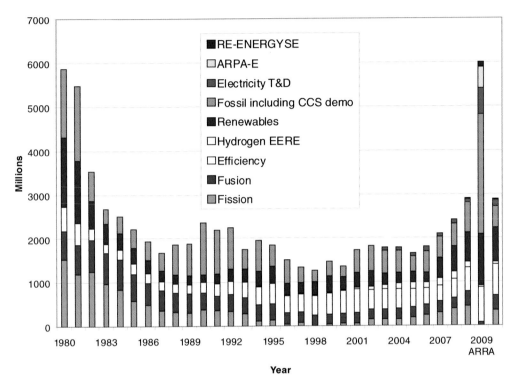

FIGURE 5.4 DOE budget authority for energy R&D: FY 1980 to FY 2010 request (in millions of year 2000 dollars). Although annual federal R&D spending for energy has begun to grow in the past few years, it is still well below its 1980 level. SOURCE: Gallagher and Anadon (2009).

Agency [IEA]) shows government energy R&D spending as a share of gross domestic product (GDP), comparing the United States and Japan from 1974 through 2008 (IEA, 2009a). Both countries increased federal energy R&D spending in response to the oil shocks of the 1970s, then decreased funding in the 1980s when the crisis subsided. The reduction in U.S. spending, however, came earlier and was larger and more prolonged than in Japan.

Since around 1990, Japan's energy R&D spending as a share of its GDP has remained at about 0.08 to 0.10 percent. In contrast, U.S. spending as a share of GDP continued to fall until about 1997, eventually leveling off at between 0.02 and 0.03 percent. It is noteworthy that, from 1992 to 2007, Japanese government spending on energy R&D also exceeded U.S. federal spending on an absolute basis (measured in 2008 prices and exchange rates), even though the GDP of Japan is about a third that of the United States. Other governments that spend a larger share of their GDP on energy R&D than

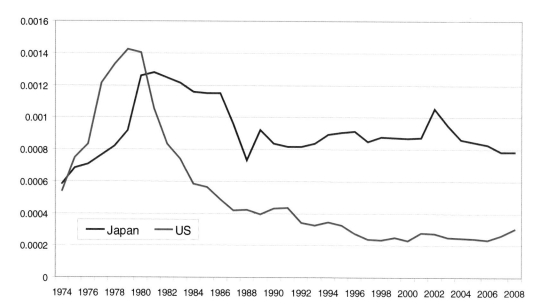

FIGURE 5.5 U.S. and Japanese government energy R&D spending as a percent of GDP: 1974-2008. For the past three decades, the U.S. percentage spending has been considerably lower than that of Japan. SOURCE: IEA (2009a).

the United States include (for the period 2005-2008) Finland, Korea, France, Canada, Denmark, Norway, and Sweden (IEA, 2009a). These data suggest that energy R&D is less of a national priority in the United States than in many other industrialized nations.

Private-Sector Funding of Energy-Related R&D

The level of private-sector funding of energy-related R&D is much more difficult to determine. The IEA estimates the total worldwide spending on energy-related R&D by private firms at between $40 and $60 billion per year, although it notes that this spending is "only partly related to clean energy" (IEA, 2009b). Although firms report R&D spending for tax purposes, they are not required to report the purpose of such spending. However, some insights are available from surveys conducted periodically by the NSF, which reports that in 2007 energy-related R&D funding by U.S. industry totaled approximately $5.3 billion.

A widely used indicator of the intensity of R&D spending by industry is the ratio of R&D spending to sales. In 2006-2007, the average ratio for all U.S.-based companies (in the top 1,400 global R&D performers) was 4.5 percent, while firms in 11 research-intensive U.S. industries spent an average of 6.5 percent (Table 5.2). Four industries showed especially high percentages: pharmaceuticals and biotechnology (16.7 percent), software and computer services (10.6 percent), technology hardware and equipment (9.6 percent), and heath care equipment and services (7.8 percent).[1] For industries where R&D spending has a high probability of being energy-related—oil/gas production and oil equipment, services, and distribution—the ratio of R&D spending to sales was in the range of only 0.2 to 2.2 percent (Table 5.3). R&D spending by the top firms identified as utilities (among the top 1,400 global R&D performers) averages ~0.7 percent of sales. Note that no U.S.-based firms in the electricity production industry are included among this list (Table 5.4).

How does this compare to R&D spending by the U.S. electricity industry? The most recent information we were able to locate was from the Government Accountability Office (GAO, 1996). This study showed that, based on data collected from 80 companies of the 112 largest operating utilities, their spending for R&D was reduced from about $708 million in 1993 to about $476 million in 1996.[2] (Spending had been level in real dollars for the previous 10 years.) In 1994, utilities on average devoted about 0.3 percent of their revenues to R&D. The GAO interviewed utility R&D managers who reported that, due to deregulation, utilities were shifting the focus of their R&D from collaborative projects benefiting all utilities to proprietary R&D, and that companies were shifting from long-term advanced technology R&D (e.g., advanced gas turbine and new fuel cells) to short-term projects that would be profitable and provide a near-term competitive edge. In fact, the R&D managers at the nation's two largest utilities, Pacific Gas & Electric and Southern California Edison, said that their advanced technology R&D programs had been eliminated.

Of course, not all energy-related R&D undertaken by private industry occurs just within the energy-producing industries; indeed, U.S. utilities have historically relied upon the companies from which they purchased equipment to undertake R&D. For example, General Electric, a major supplier to the utility industry, spent $3 billion on R&D in 2007-2008, which represents 2 percent of its sales and 12 percent of its profits. The company was ranked 17th in the United States and 43rd in the world in terms of R&D spending. However, because of the complexity of corporate structures and business

---

[1] These figures are based on data for total R&D spending by the top 1,400 firms in the world, of which 536 firms were based in the United States (DIUS, 2009). To be included in this list, a firm had to spend at least $36 million in R&D. To avoid distortion due to a small number of firms in an industry, the data in Table 5.2 include only industries in which there are 10 or more U.S.-based firms included in the list.

[2] The 112 largest investor-owned public utilities accounted for over 93 percent of all nonfederal utility R&D spending and were responsible for about three-quarters of all electricity sales.

TABLE 5.2 Year 2007-2008 R&D Spending by Industry for U.S.-Based Firms Included in the List of the Top Global Companies Ranked by R&D Investment

	R&D Investment (millions)	Sales (millions)	R&D/Sales (%)
All U.S.-based companies composite (536 companies)	$209,026	$4,622,394	4.5
Industry sectors reporting data for 10 or more U.S.-based firms (number of U.S. companies)			
Pharmaceuticals and biotechnology (80)	$50,410	$302,159	16.7
Software and computer services (72)	$29,712	$279,566	10.6
Technology hardware and equipment (134)	$51,833	$540,022	9.6
Health care equipment and services (36)	$6,831	$87,093	7.8
Automobiles and parts (19)	$20,617	$529,056	3.9
Electronic and electrical equipment (34)	$4,204	$111,187	3.8
Aerospace and defense (17)	$9,553	$287,064	3.3
Chemicals (28)	$5,949	$222,893	2.7
Household goods (11)	$3,227	$131,401	2.5
Industrial engineering (20)	$4,411	$182,367	2.4
General industrials (13)	$6,745	$284,779	2.4
Total these sectors (464 companies)	$193,492	$2,957,588	6.5
Percent of all U.S.-based companies for which data are reported (86.6%)	92.60%	64.00%	

NOTE: Includes only industries listing 10 or more U.S. firms, and R&D spending is worldwide spending by these U.S.-based companies. SOURCE: Compiled from the U.K. Department for Innovation, Universities & Skills, the 2008 R&D Scoreboard (2009).

TABLE 5.3 R&D Investment by U.S. Firms in the Oil and Gas Production and the Oil Equipment, Services, and Distribution Industries Included in the 2008 List of the Top 1,400 Global Companies Ranked by R&D Investment

	R&D Investment (million)	Sales (million)	R&D/Sales (%)
Oil and gas production[a]	$1,536	$741,138	0.2
Oil equipment, services, and distribution[b]	$1,905	$85,009	2.2

[a]Exxon Mobil, Chevron, and ConocoPhillips

[b]Slumberger, Baker Hughes, Halliburton Weatherford International, Smith International, BJ Services, FMC Technologies, Grand Prideco, and McDermott International.

SOURCE: Same as Table 5.2.

TABLE 5.4 R&D Spending by All Firms Identified as "Utilities" in the 2008 List of the Top 1,400 Global Companies Ranked by R&D Investment.

	R&D (million)	Sales (million)	R&D/Sales (%)
Total all firms	$2,670	$413,416	0.6
Korea Electric Power, South Korea	$644	$31,127	2.1
Electricité de France, France	$548	$87,192	0.6
Vattenfall, Sweden	$174	$22,225	0.8
Hydro-Quebec, Canada	$101	$12,497	0.8
Iberdrola, Spain	$95	$25,539	0.4
Taiwan Power, Taiwan	$64	$12,391	0.5
Energie Baden, Germany	$58	$21,652	0.3
Enel, Italy	$42	$62,423	0.1
Total for eight Japanese firms[a]	$942	$138,369	0.7

[a]Tokyo Electric, Kansai Electric, Chibu Electric, Khushu Electric, Tohoku Electric, Chugoku Electric, Electric Power Research, and Shikoku Electric.

SOURCE: Same as Table 5.2.

units, there is no clear way to determine the energy-related portion of that total R&D spending. GAO cites an estimate (from a study of the Electric Power Research Institute) that in 1988 all U.S. manufacturers spent $200 million on electricity-related R&D (GAO, 1996).

Overall, these data suggest that the current rate of R&D spending by U.S. energy industries is far below that of other industries whose products and profits depend more strongly on the ability to innovate. This suggests that transforming the U.S. energy sector will require a significant increase in private-sector R&D spending to develop and commercialize new low-emissions technologies. In general, the level of private-sector spending on R&D is motivated mainly by its value to a firm's profitability. Thus, substantial increases in energy-related R&D expenditures will occur only if government policies create conditions under which firms anticipate that such spending is likely to yield attractive financial returns in the foreseeable future.

Workforce Available to Support Technological Innovation

Accelerating the pace of technological change to reduce GHG emissions will require a skilled workforce in all stages of the technological change process depicted in Figure 5.2 (i.e., not only for the development but also for the diffusion of new technologies). Chapter 6 discusses the overall employment impacts of policies to reduce GHG emissions. This section focuses on one of the key human resources needed to support a vigorous, sustained effort in technology innovation: the supply of scientists and engineers working in industry, universities, government laboratories, and other venues to invent and bring to commercial readiness a host of new technologies.

R&D Scientists and Engineers in Industry

The NSF publishes information on the number of R&D scientists and engineers employed by various industries. Table 5.5 shows these data for 2004, the latest year for which they are available. In that year, industry employed a total of 1.1 million R&D scientists and engineers. Seventy-five percent were employed by only five industries: computer and electronic products; professional, scientific, and technical services; transportation equipment; information (including telecommunications and software); and chemicals. All of these five industries (except transportation) had more than 100 R&D workers per 1,000 employees, or more than 10 percent of the workforce. For the computer and electronic products industry, 20 percent of the workforce was in R&D. The average across all industries was 7.1 percent.

For utilities (the only energy-producing industry for which NSF provides data), R&D employees constitute only 0.3 percent of the labor force—a very low level relative to other U.S. industries. This is consistent with the comparatively low level of R&D funding seen earlier. To match the average R&D workforce for all U.S. industries, the utilities industry alone would need to employ an additional 19,000 R&D scientists and engineers.

As noted earlier, R&D relevant to the utility industry also is performed in other industries that service and supply technology to utility firms. Unfortunately, however, data are not readily available to obtain a comprehensive picture of this broader energy R&D workforce, or of other segments of the economy responsible for GHG emissions. Nonetheless, the qualitative conclusion that emerges from available data is that the energy sector does not appear well positioned at present to undertake a major program of technological innovation to reduce GHG emissions. Additional personnel with science and engineering training will be needed.

Engineering and Science Graduates

The NSF publishes data on the number of graduates from U.S. colleges and universities majoring in different disciplines (NSF, 2008). Figure 5.6 shows the trends in undergraduate (B.S.) and graduate (M.S. and Ph.D.) degrees in fields of science and engineering (S&E)[3] and compares this to the trends in total non-S&E degrees over this period. All data are normalized to the total U.S. population in a given year to give a better picture of the relative popularity of different degree programs. Undergraduate degrees in non-S&E fields (as a percentage of the U.S. population) have risen significantly since 1985. In contrast, the percent of undergraduate degrees in all S&E fields has been relatively flat over the past four decades. The trend in total graduate degrees (M.S. plus Ph.D.) over the past two decades for S&E fields has been roughly constant since 1970, while for non-S&E fields there was steady and sizeable growth. Figure 5.7 shows further details for the S&E category, displaying the separate degree trends for science and engineering fields. While the percentage of B.S. degrees in science fields has grown significantly over the past two decades, the percentage for engineering fields has declined by nearly 50 percent since its high in the mid-1980s.

Sustained progress in technological innovations to reduce GHG emissions is likely to require increasing numbers of engineers and scientists in a variety of disciplines. This will require reversing some of the historical trends displayed in Figure 5.5. Actions by both the private sector and the public sector will be needed to achieve that end. Direct and indirect support for technological innovation also will require the training of large numbers of industrial workers who may not require a college degree but who

---

[3] The S&E fields included here are biological and agricultural sciences; earth, atmospheric, and ocean sciences; mathematics and computer sciences; physical sciences; and engineering. See NSF (2008) for details of the subfields included in each category. The non-S&E figures include psychology and social sciences.

TABLE 5.5 Employment of R&D Scientists and Engineers for Companies, Listed by Industry, That Perform Industrial R&D in the United States

Industry	R&D Scientists and Engineers (thousands)	R&D Scientists and Engineers as Share of Total Domestic Employment (%)
Software	93.7	39.4
Scientific R&D services	44.7	27.4
Architectural, engineering, and related services	41.4	26.4
Communications equipment	49.9	23.8
Semiconductor and other electronic components	97.4	23.7
Computers and peripheral equipment	45.1	18.3
Pharmaceuticals and medicines	79.9	17.0
Navigational, measuring, electromedical, and control instruments	74.6	16.6
Computer systems design and related services	74.5	15.4
Other information	22.0	11.5
Other computer and electronic products	6.2	11.3
Wholesale trade	15.5	10.0
Machinery	62.6	9.4
Resin, synthetic rubber, fibers, and filament	9.4	9.4
Other professional, scientific, and technical services	13.5	8.9
Motor vehicles, trailers, and parts	89.2	8.6

will need advanced (post-high school) training offered in community colleges and other institutions.[4]

---

[4] The American Wind Energy Association reported that, at the end of 2008, it was able to identify over 100 educational institutions that are offering or developing programs that focus on wind or renewable energy. These programs ranged from certificate programs and 2-year associate degrees that focus on wind technician training, to bachelor's and graduate degrees that benefit a range of areas specific to the wind industry (AWEA, 2009).

TABLE 5.5 Continued

Industry	R&D Scientists and Engineers (thousands)	R&D Scientists and Engineers as Share of Total Domestic Employment (%)
Medical equipment and supplies	13.9	6.6
Aerospace products and parts	37.9	6.1
Basic chemicals	10.6	5.9
Other chemicals	18.6	5.7
Electrical equipment, appliances, and components	19.4	5.6
Other miscellaneous manufacturing	7.9	5.5
Newspaper, periodical, book, and database	4.8	4.6
Health care services	6.0	3.8
Plastics and rubber products	14.1	3.3
Fabricated metal products	15.7	3.3
Paper, printing, and support activities	14.6	3.1
Finance, insurance, and real estate	22.3	2.6
Other transportation equipment	7.0	2.4
Broadcasting and telecommunications	10.9	1.6
Food	11.7	1.3
Construction	0.8	0.5

NOTES: Data recorded on March 12, 2004, represent employment figures for the current year. Data recorded in January 2005 represent employment figures for the previous year. The method used to assign industry classifications has changed; industry-specific estimates for 2004 are not directly comparable with previous years. Detail may not add to total because of rounding. Excludes data for federally funded R&D.
SOURCE: National Science Foundation Division of Science Resources Statistics, Survey of Industrial Research and Development: 2004.

Key Findings on Available Resources

Fostering technological innovations to reduce GHG emissions on a large scale will require widespread efforts to develop and deploy low-emissions technologies. A vibrant and sustained R&D program, especially focused on low-emissions energy production and utilization technologies, is a cornerstone of that process, along with the availability of a skilled R&D workforce. In this regard, a review of current U.S. resources devoted to such efforts, in both public and private sectors, presents cause for concern.

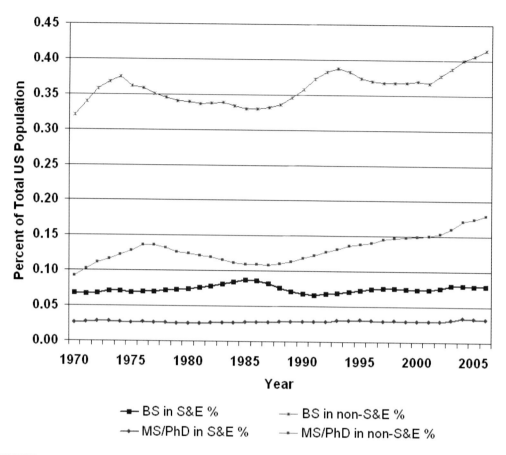

FIGURE 5.6 Trend in U.S. undergraduate (B.S.) and graduate (M.S. plus Ph.D.) degrees granted as a percentage of the total population. The science and engineering (S&E) fields include biological and agricultural sciences; earth, atmospheric, and ocean sciences; mathematics and computer sciences; physical sciences; and engineering. Non-S&E includes all other degrees. The trend in total graduate and undergraduate degrees in S&E as a percentage of the U.S. population has been roughly constant since 1970. SOURCES: Based on data from NSF (2008); U.S. population data from U.S. Census Bureau (2000, 2007).

Federal Support for Energy-Related R&D

Federal funding for energy-related R&D has declined dramatically over the past several decades, both in absolute terms and relative to other national R&D priorities such as health and space exploration. The United States also lags behind other leading countries in the fraction of national resources devoted to energy-related R&D. To achieve parity with health-related federal R&D spending (as was the case in 1980), energy-related R&D would have to be increased 20-fold over recent (FY 2008) levels—an increase of $23 billion per year (in constant FY 2000 dollars). To achieve parity with

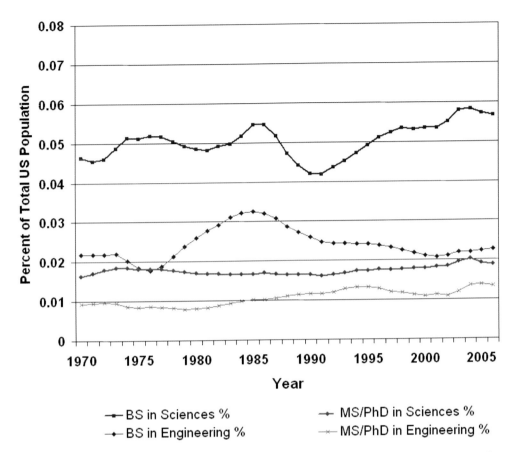

FIGURE 5.7 Breakdown of S&E trend in U.S. undergraduate (B.S.) and graduate (M.S. plus Ph.D.) degrees granted as a percentage of the total population, showing the separate trends for science fields and engineering fields. The trend in undergraduate science degrees as a percentage of the U.S. population has been rising over the past two decades but the percentage for undergraduate engineering degrees has fallen significantly since the 1980s. SOURCE: Based on data from NSF (2008).

federal funding for space-related R&D would require a sixfold increase—roughly an added $7 billion per year. Even a return to 1980 levels of energy-related R&D would require an additional $5 billion per year in federal funding relative to FY 2008 levels.

Although such increases are large relative to recent budgets, they are likely to represent only a portion of the total (government plus private sector) R&D investment needed in the decades ahead. It is important to keep in mind, however, that the potential returns from R&D are very large. For example, Figure 5.1 suggests that having more advanced technology choices available to control GHG emissions could greatly reduce the carbon price needed to control emissions. If the United States is to reduce its GHG emissions to attain the budget targets discussed in Chapter 2, even a small reduction in future carbon prices as a result of R&D investments would significantly reduce the total cost of achieving these targets.

This report does not attempt to recommend the structure or priorities of a federally supported R&D agenda to limit the magnitude of climate change, because detailed assessments of such R&D needs have been developed in many other recent studies by the National Research Council and other organizations (e.g., American Physical Society, 2008; DOE, 2009; NRC, 2009a; PCAST, 2008; IEA Technology Roadmaps[5]). Those efforts represent a strong starting point for further deliberations and planning. What is emphasized in this report is the urgent requirement for substantially increased federal resources for R&D to reduce GHG emissions. Substantial additional R&D funding from the private sector also is critical, and federal policies will be needed to bring this about.

Although federal R&D funding cannot by itself ensure that energy-saving and GHG-reducing technologies will achieve widespread use in an economy, it nonetheless plays a critical role in the overall process of technological change. Federally funded basic R&D provides the starting point for many (if not most) significant energy-related innovations, and federally funded assistance for technology development often is the catalyst for turning technological innovations into practical products that are sought in the marketplace (NRC, 2001). And while the *level* of federally funded R&D is extremely important, it is also important that funding be *relatively stable* over time—not only to attract and retain a high-quality R&D workforce but also to avoid the disruptive "boom and bust" patterns of past federal energy R&D spending, which causes the private sector to question the long-term market potential and value of pursuing new product development.

Private-Sector Support for Energy-Related R&D

Private-sector funding of energy-related R&D is also critical for achieving the innovations needed to reduce GHG emissions on a large scale. Here too, however, the current picture for U.S. industries appears rather bleak. A widely used indicator of innovative activity is the ratio of R&D spending to industry-wide sales. For U.S.-based energy companies, this ratio is far below that of other leading technology-based industries, suggesting a major shortfall in the ability of the U.S. energy industry to bring about the technological innovations that are needed. Transforming the U.S. energy industry

---

[5] For instance, see *http://www.iea.org/subjectqueries/keyresult.asp?KEYWORD_ID=4156.*

in a carbon-constrained world will clearly demand a significant increase in private-sector R&D.

Workforce to Support Technological Innovation

Achieving major innovations to reduce GHG emissions will require a workforce with a broad range of skills. The ability to innovate will especially require increasing numbers of engineers and scientists in a variety of disciplines. As a percentage of the total workforce, however, limited data suggest that U.S. energy industries currently have far fewer R&D workers than the average for all U.S. industries. The percentage of U.S. graduates in engineering also has declined significantly in the past two decades. Actions by both private and public sectors will be needed to reverse these trends.

ASSESSMENT OF CURRENT U.S. INNOVATION POLICIES

As noted earlier, achieving and accelerating technological changes that reduce GHG emissions on a significant scale will require policies not only to promote and sustain a vigorous program of R&D but also to establish and expand markets for low-GHG technologies and to help commercialize and bring new technologies to the marketplace. Table 5.1 listed some of the "technology policy" options available, along with regulatory policies, to help achieve these ends. This section briefly assesses the actual state of current U.S. policies related to these objectives to identify any major deficiencies or limitations that must be addressed to foster needed technological innovations.

Overview of Current Policies

The federal government has many programs, policies, and measures aimed at encouraging the commercialization and deployment of technologies that reduce, avoid, or capture and sequester emissions of GHGs. A recent report by the Committee on Climate Change Science and Technology Integration (CCCSTI) (DOE, 2009) identified more than 300 such policies and measures, which it grouped into nine categories of "deployment activities": tax policy and other financial incentives; technology demonstrations; codes and standards; coalitions and partnerships; international cooperation; market conditioning including government procurement; education, labeling, and information dissemination; legislative acts of regulation; and risk mitigation.[6] The

---

[6] The CCCSTI was a cabinet-level committee created in February 2002 to coordinate climate change science and technology research. At the time the above report was released, the Secretary of Energy was

comprehensive taxonomy and lists developed in the CCCSTI report underscore just how many different federal policies can impact innovation related to the reduction of GHGs, and the importance of choosing policies that will have the intended impacts.

Current U.S. policies that directly or indirectly support technological innovation have contributed to the progress made to date in decreasing the demand for some types of energy services (such as lighting and refrigeration) and in enabling increased use of some low-carbon energy supplies (such as wind and solar power). However, there are critical areas in which current policies and actions are inadequate or absent.

Limitations of Current U.S. Policies

As catalogued in detail in the CCCSTI report, current U.S. policies have gaps in certain technology areas or market sectors that are important to reducing GHG emissions. Many of these policy gaps result in a lack of clear market signals regarding commercial opportunities for companies to invest in new technologies and related R&D to enhance their competitive advantage. Table 5.6 summarizes an array of policies deemed by Brown and Chandler (2008) as unfavorable (those that place clean energy technologies at a competitive disadvantage) or as ineffective (those with design flaws that undermine their intended outcomes).

At present, market incentives for low-GHG technologies are driven primarily by state- or regional-level policies, such as renewable-energy portfolio standards, with federal R&D assistance at the earlier stages of development, as previously discussed. Without such regulatory requirements or an explicit price on GHG emissions, however, low-GHG technologies must compete in markets that do not value low emissions. Thus, the most critical need at this time is for federal policies that create or expand markets for low-GHG emissions technologies on a level playing field. As discussed in Chapter 4, we suggest that a portfolio of strategies, including an appropriate price on GHG emissions together with strategically targeted complementary policies, is the most effective way to foster innovation and markets for new technologies.[7]

The existence of policies does not imply or guarantee their effectiveness. For example,

---

Chair of the CCCSTI, the Secretary of Commerce was Vice Chair, and the Director of the White House Office of Science and Technology Policy was the Executive Director.

[7] We note that market dynamics may not prove effective in certain cases. For example, the regulated power supply industry faces limited opportunities to garner returns on R&D investments because public utility commissions rule on rate cases and set prices and returns on investments that are generally less than the profits sought in competitive industries. Thus, additional incentives or mechanisms may be needed to achieve the desired outcomes.

TABLE 5.6 Examples of Policy Impediments to Clean Energy Technologies

Unfavorable Fiscal Policies	Tax subsidies	Oil and gas depletion allowances allow owners to claim a depletion deduction for loss of their reserves.
		The link between federal transportation funding and vehicle miles traveled rewards the growth of transportation energy use.
	Unequal taxation of capital and operating expenses	The federal tax code discourages capital investments in general, as opposed to direct expensing of energy costs.[a]
	Unfavorable tariffs	Utilities impose tariffs (e.g., standby charges, buyback rates, and uplift fees) on small generators seeking to connect to the grid.[b]
	Utility pricing policies	Unfavorable electricity pricing policies present obstacles for an array of clean energy technologies; these include the regulated rate structure, lack of real-time pricing, and imbalance penalties.
		In traditionally regulated electricity markets, electric utilities face little incentive to promote energy efficiency or distributed generation, because utility company profits are a function of sales. California and a few other states have decoupled utility profits from energy sales. Fourteen states have enacted decoupling in natural gas markets and six states have in electricity markets.
Ineffective Fiscal Policies	Examples	The Internal Revenue Service has yet to establish guidelines that clarify the eligibility criteria and spell out procedures for claiming tax credit for fuel cells authorized in the Energy Policy Act of 2005.
		Tax credits intended to promote the purchase of hybrid electric vehicles and residential photovoltaic systems have limited value because of the Alternative Minimum Tax, which sets a floor for tax liability.
		Many states have property tax laws that provide incentives for landowners to develop their forestland for higher use rather than leave the forest standing or continue timber production.

Continued

159

TABLE 5.6 Continued

Fiscal Uncertainty	Fiscal incentives	Limited-duration tax policies such as production or investment tax credits.
	Fiscal penalties	Utilities may be penalized for promoting energy efficiency due to reduced sales, if the energy efficiency program and impacts are not accounted for in rate rulings or other fiscal measures (e.g., return on assets vs. sales).
Unfavorable Regulatory Policies	Performance standards	Exempting existing facilities from strict emissions requirements placed on new plants discourages technological progress and capital stock turnover.
		Emissions standards that are input based rather than output based discourage process improvements that would result in lower emissions.
	Connection standards	The ban on private electric wires crossing public streets penalizes local generation of electricity, which could reduce transmission losses and increase overall efficiency.[c]
Ineffective Regulations	Regulatory loopholes	Federal Corporate Average Fuel Economy standards credit vehicles for flexible fuel (E-85 capability) regardless of how they are fueled after purchase or their fuel mileage.
		Zoning for low-density urban development contributes to sprawl and locks in dependence on cars rather than multiuser transit.
		Several clean energy technologies, such as carbon capture and hydrogen, are challenged by inadequate regulatory frameworks.
	Burdensome permitting processes	Regulatory uncertainty—regarding whether or not GHG will be regulated or how current technologies will fare under new regulatory processes—impedes rational investment decisions.
		Multiple agency reviews and approvals required for most energy facilities slow down the process and place an undue burden on the developer.

TABLE 5.6 Continued

Regulatory Uncertainty	Lack of modern and enforceable building codes	Building codes that are not enforced are based on outdated technology or allow trade-offs that mitigate use of existing technology discourage adoption of clean energy technologies.
Unfavorable Statutes	State procurement policies	When state agencies cannot contract over more than one fiscal year, they are unable to take on capital improvements that are cost-effective in the long run.[d]
	Uncertain property rights	Property rights for subsurface and above-surface areas are unclear. In some cases, particularly coalbed methane, geologic storage of carbon dioxide, and wind energy, property rights for these areas must be defined to provide investment certainty.

[a] U.S. tax rules require capital costs for commercial buildings and other investments to be depreciated over many years, whereas operating costs can be fully deducted from taxable income (26 USC § 168). Since efficient technologies typically cost more than standard equipment, this tax code penalizes efficiency.

[b] For example, small generators hoping to connect to the grid in the mid-Atlantic area must undergo a review at a cost of $10,000 to the generator before being allowed to tap into the PJM (Pennsylvania, New Jersey, Maryland) grid interconnection (Sovacool and Hirsch, 2007)

[c] Bans on private wires and metering rules have historically inhibited the installation of distributed generation (DG) systems in the United States (Alderfer et al., 2000; Mueller, 2006; Sovacool and Hirsh, 2007).

[d] Energy service companies deliver energy-efficiency upgrades to industrial, commercial, and government facilities through energy-saving performance contracts with an energy service company—a business that develops, installs, and finances projects to improve the energy efficiency and maintenance costs of facilities. Increasingly, this contracting mechanism is being used by government agencies to upgrade the efficiency of government-owned buildings. But many state constitutions prohibit the obligation of funds in advance of their being appropriated, which can prohibit multiyear contracting with energy services companies. SOURCE: Based on Brown and Chandler (2008).

the production tax credit is often cited as a key policy instrument for stimulating investments in clean energy technologies, such as wind turbines. However, in the United States there is a history of enacting such policies for relatively short durations, followed by reauthorization after the policy has expired. This "on again, off again" behavior creates strong market uncertainty and causes abrupt changes in business investments and R&D (Wiser et al., 2007). Thus, stability in the policy environment is an important factor in sustaining a climate of technology innovation. This is essential to encourage consistent corporate investments over the long lead times needed for commercial success.

As noted earlier, U.S. government investments in energy-related R&D programs have been very low compared to R&D in other areas such as health and space. Private-sector R&D in energy industries also is extremely low compared to the national average for all industries, and especially low compared to innovative industries such as information technologies or biotechnology. Expanding direct R&D programs for low-emissions energy production and utilization, as well as other areas related to GHG emissions reductions, is also a critical element of a comprehensive policy.

As seen earlier, current levels of R&D workers in energy industries also are significantly below industry-wide averages, and the downward trend in percentages of U.S. graduates in key professional fields such as engineering is not encouraging. Thus, expanded education and workforce development efforts are likewise crucial elements of any innovation strategy. There is a need for increased financial support across the educational spectrum. Alignment of program strategies in research organizations and government agencies that support education and workforce development—including but not limited to NSF, the U.S. Department of Agriculture, DOE, and the Departments of Interior, State, and Defense—may offer significant synergies and leverage in this regard. Greater efforts to not only attract but also retain innovation scientists would help address the trend of foreign nationals obtaining advanced degrees in the United States and then emigrating for future employment.

Even with new innovations, the pace of technological change can be slowed by the prevalence of long-lived assets (especially in the energy sector) and the potential for technology lock-in (Grübler, 2004). Policy approaches necessary to transform the landscape of long-lived assets at the appropriate speed and scale are discussed in Chapter 4.

As described earlier, residential and commercial buildings represent one of the key sectors with major opportunities for innovations that help limit GHG emissions. Institutional barriers such as the "principal agent" problem have been reported in detail (Popp et al., 2009) and were discussed in Chapter 4 of this report. Alignment of incentives to support technology innovation, together with improved community development planning in general, would enable significant reductions in U.S. energy demand and associated GHG emissions, with high effective returns (e.g., low abatement costs). Policies to support early commercialization and large-scale adoption are also lacking in a number of critical areas, such as carbon sequestration and non-CO_2 GHG reductions. These deficiencies must be dealt with effectively through the range of policy measures outlined above and elaborated earlier in Chapter 4.

The items highlighted above are not intended to be a comprehensive list of the limitations in current U.S. policies affecting technological innovation. They do, however, represent what we believe are some of the most crucial issues that must be addressed to create an effective policy environment for innovations needed to reduce GHG emissions. Many of these items also have been identified in other recent studies (e.g., NSB, 2009). The following section presents a number of recommendations based on the findings and conclusions of this chapter.

KEY CONCLUSIONS AND RECOMMENDATIONS

A critical component of any climate change limiting policy is technological change that replaces current GHG-intensive technologies with low-GHG technologies. In many cases this requires advanced technologies that have not yet been invented, commercially developed, or adopted at a significant commercial scale. Given the magnitude and rate of technological changes needed to limit future climate change, there is an immediate need to adopt policies to accelerate technological innovation throughout the U.S. economy. Such policies also will enhance the competitive position of the United States as a developer and marketer of technologies to address climate change.

The process of technological change involves the stages of invention, innovation, adoption, and diffusion. The first two stages encompass the traditional domain of R&D, while the latter stages require markets for new or advanced technologies. Because these stages are highly interdependent, policies to promote technological innovation need to be comprehensive and not focused solely in one area.

Direct federal support for R&D and the training of a skilled R&D workforce are especially critical for fostering technological innovation. The Federal Budget is a powerful and tangible statement of the nation's priorities. Comparing across different policy areas, we find that federal R&D spending on energy in FY 2008 was approximately one-twentieth of federal R&D spending on health, one-sixth of federal R&D spending on space, and one-fifth of federal R&D spending on general science. Comparing across time, we find that energy R&D spending in FY 2008 accounted for approximately 2.6 percent of total federal (nondefense) R&D spending, a 10-fold decline from its peak of approximately 25 percent in FY 1980. Comparing internationally, we find that U.S. spending on energy R&D as a share of GDP is considerably lower than that of several other leading industrialized countries.

The 2009 American Recovery and Reinvestment Act provided a significant one-time increase in federal energy R&D expenditures, but we have not yet seen the type of sustained changes in federal R&D spending that would indicate energy to be a high national priority. While recommendations for desired levels and priorities for federal energy R&D spending are outside the scope of this study, we do find that the level and stability of current spending do not appear to be consistent with the magnitude of R&D resources needed to address the challenges of limiting climate change. In regard to the private sector, compared to other U.S. industries, the U.S. energy sector currently spends very little on R&D relative to income from sales and employs very few R&D workers. Substantial increases in private-sector investments in R&D will be needed to foster innovation and technological change.

Creating and expanding markets for low-GHG-emissions technologies are also critical to spur innovation. Table 5.1 identifies a wide range of policy options, in addition to traditional R&D spending, that the federal government can use to help spur techno-logical innovation and expand markets for low-GHG-emissions technologies. Creating some form of substantial and sustained carbon-pricing system (discussed in Chapter 4) is likely to be of the utmost importance in stimulating the development and de-ployment of new technologies and approaches to reducing GHG emissions.

CHAPTER SIX

Interaction with Other Major Policy Concerns

The issue of limiting future climate change cannot be addressed in isolation; it is closely linked with many other issues of major public interest and concern. While it is beyond the scope of this study to explore the full range of complex mutual dependencies among environment, energy security, economy, and societal well-being (i.e., the broader concepts of sustainable development), in this chapter we do explore a few key issues that influence or are influenced by climate change limiting policies. We first consider some areas where policies for limiting climate change can potentially offer added benefits on domestic and international levels, including energy security, protection of air and water quality, and adaptation to the expected impacts of climate change. We then explore how our nation's response to climate change affects existing concerns about equity and environmental justice, including the question of how employment opportunities across the United States may be affected.

LINKAGES WITH ENERGY AND ENVIRONMENTAL POLICY ISSUES

Policies to limit climate change may enhance or detract from the effectiveness of policies designed to achieve other national goals. For example, technology for controlling CO_2 emissions from the electricity and transportation sectors may also reduce emissions of other pollutants that result from the combustion of fossil fuels. Similarly, reducing the consumption of conventional petroleum to diminish CO_2 emissions may also minimize the nation's vulnerability to oil price shocks. And a CO_2-emitting company might meet its emissions targets by paying a farmer to sequester carbon using agricultural practices that simultaneously reduce erosion and water pollution.

Existing research describes how these ancillary benefits might arise and, in some cases, estimates their possible magnitude. This section briefly reviews such research, considering in particular how policies designed to limit future climate change could enhance energy security, reduce air pollution, and mitigate adverse effects of agriculture and forestry practices. In addition, we explore the synergies, both domestic and international, between strategies to limit climate change and strategies to adapt to the consequences of climate change. Based on this brief review, the panel suggests how U.S. climate change limitation policy might reasonably be advanced by taking advantage of ancillary benefits in these areas.

Domestic Ancillary Benefits and Costs

Energy Security

Reducing oil use in the transportation sector can be achieved through greater efficiency, substitution of noncarbon fuels, and electrification of transport systems.[1] Opportunities also exist to reduce oil consumption in the buildings sector (home weatherization to reduce heating oil usage) and the industrial sector (increasing the efficiency of industrial processes that use liquid fuels). Reducing oil consumption would not only help reduce our nation's GHG emissions but also have the benefit of minimizing the nation's economic vulnerability to high oil prices and potential supply disruptions, thus enhancing energy security.

Reducing oil consumption has two effects on energy security. First, because the United States consumes a quarter of world oil production, modest changes in U.S. demand can substantially affect the supply-and-demand balance of world markets. Thus, U.S. reduction in oil consumption should result in a lower world oil price. If it does, U.S. expenditures for petroleum will drop because both the quantity of oil used and the price per barrel drops. This change is the economic value of the ancillary benefit.

The second effect of reducing oil consumption is to buffer the economic effect of potential oil price shocks due to natural- or human-caused supply interruptions. Such interruptions appear to be a real possibility. Huntington (2008) interviewed experts in oil markets and concluded that there is a 50 percent chance of a disruption of two million barrels of oil per day lasting at least 30 days. A disruption of this magnitude would create a spike in oil prices and produce other stresses in the economy. The size of these effects would be directly proportional to the nation's dependence on oil as an energy source, and so reducing that dependence creates an economic value.

Leiby (2007) has estimated the economic value of these two factors,[2] and Huntington (2008) and Parry et al. (2007) have reviewed this estimating methodology. They point out that the uncertainties are large and that the estimated values are sensitive

---

[1] As discussed in Chapter 3, projections (reaching until ~2030) indicate that the GHG emissions reductions achieved through these strategies will likely be offset by the growing demand for travel services. Thus, strategies for reducing travel demand are needed as well.

[2] Leiby's estimates do not include terms of trade effects, nor do they include costs that are hard to allocate, like the cost of maintaining a military presence in oil-supplying regions.

to the prevailing price level and to judgments about the risks of disruption and the economic response to a disruption. Nevertheless, these sources agree that Leiby's estimates are reasonable representations of the costs of U.S. dependence on oil. Furthermore, the National Research Council (NRC, 2002b) used Leiby's work as the basis for estimating the side benefits to energy security associated with more stringent fuel economy standards.

These sources estimate that the value of reduced oil consumption (in 2007 dollars) averages about 15 cents per gallon of gasoline,[3] although the range of estimated values is broad. For example, Leiby (2007) estimates a range of 10-30 cents per gallon, and Parry et al. (2007) suggests a range of 8-50 cents per gallon. In addition, the estimates could change substantially with variations in oil prices and average fuel economy of the automobile fleet. In particular, as oil prices drop, the value of the ancillary benefit drops as well.

Reducing the use of coal and natural gas does not affect energy security because they are supplied almost entirely from domestic sources and neither presently displaces oil in the transportation sector. Furthermore, the electric power sector uses very little oil and so does not affect domestic economic vulnerability to oil dependence. Note that some actions taken to enhance energy security can actually exacerbate climate change. For example, if Canadian oil sands or coal-based syn-fuels were to substitute for imported petroleum, greenhouse gas (GHG) emissions could rise because of the energy-intensive production processes involved.

Air Pollution

In addition to creating CO_2, the combustion of fossil fuels produces a variety of pollutants controlled under the Clean Air Act, including particulate matter (PM), nitrogen oxides (NO_X), volatile organic compounds (VOCs), and carbon monoxide (CO), which contribute to the formation of photochemical smog that has adverse human health effects. Coal combustion in the electric sector produces sulfur oxides (SO_X), NO_X, and PM that not only injure human health but also damage vegetation and acidify lakes and streams. The process of coal mining is also a source of considerable damage to the environment and human health (e.g., Palmer et al., 2010).

Emissions of these pollutants have been significantly reduced since the implementation of the Clean Air Act in the early 1970s. Nevertheless, some emissions remain and continue to create adverse effects for human health and natural ecosystems. NRC

---

[3] One cent per gallon of gasoline is roughly equivalent to one dollar per ton of CO_2-eq.

(2009c) and Parry et al. (2007) estimated residual pollution damages to health and other effects and, by applying the value of a statistical life used in Environmental Protection Agency (EPA) cost-benefit analyses, they assigned an economic value to these estimated health damages. In the transportation sector, health-related damages arise from emissions of various compounds, including PM, NO_x, SO_x, and VOCs. The health damage estimates from NRC and Parry et al. are in the general range of 29 to 40 cents per gallon for conventional gasoline-fueled internal combustion engines (although estimates vary somewhat depending on technology and exposure assumptions). In the electricity-generation sector, the NRC estimates the mean residual air pollution damages at 3.2 cents per kilowatt hour (weighted by net generation and expressed in year 2007 dollars) for coal-fired power plants, but the range of damages is very large. The NRC calculated damages for each of 406 coal-fired power plants and found that damages from plants with full modern pollution control technology can be one-tenth or less of the mean estimate. In contrast, inefficient plants with less effective pollution controls could have damages about four times greater than the mean. Thus, the residual damages can be significant, but they are highly dependent on factors such as plant vintage, sulfur content of coal burned, type of emissions controls, and proximity to population centers.

Fossil fuel PM emissions include black carbon particles, which not only have adverse human health effects but also exert strong but positive radiative forcing (i.e., warming) in the atmosphere. Mitigating black carbon emissions therefore offers both health and climate change benefits. In contrast, the sulfate aerosols generated by fossil fuel combustion exert negative radiative forcing (i.e., cooling) by reflecting solar radiation; thus reducing these emissions for the purpose of mitigating health impacts can actually exacerbate climate change (discussed further in Chapter 2).

Agriculture and Forestry

A carbon-pricing system may include the option for the GHG emitter to purchase (domestic or international) offsets from the agricultural or forestry sectors, if the offset purchase were cheaper than the cost of controlling the GHG-emitting source. Offset activities may include increasing soil and ecosystem sequestration of carbon (e.g., through minimum tillage practices, halting timber harvesting, reducing emissions from deforestation and forest degradation [REDD] programs, or preventing fires) and reducing agricultural sector emissions by limiting fertilizer use, improving manure management, or reducing livestock herd size. These activities could in turn produce important ancillary benefits such as cleaner water and less soil erosion. Elbakidze and McCarl (2007) estimated that these ancillary benefits can be approximately $1-3 per ton of CO_2. They note, however, that the use of offsets allows the primary GHG emitter to continue with more fossil fuel emissions, leading to ancillary health costs, which are on the same order of magnitude as the value of the ancillary benefits gained from agricultural sequestration. Emissions leakage may also be a concern, as discussed in Chapter 4.

Climate Change Adaptation

Limiting future climate change and adapting to whatever degree of climate change occurs both involve costs. In principle, these costs could be traded off against one another. If adapting to climate change were less costly than taking action to prevent that change, then policy makers might prefer to let the change happen. As discussed elsewhere in the report, reduction in U.S. GHG emissions, in isolation, will have little *direct* marginal effect on future climate change; however, U.S. emissions-reduction efforts could have a large *indirect* influence on climate outcomes by affecting how other countries act. It is thus very difficult to assess whether more aggressive policies to limit U.S. GHG emissions would lead to any marginal benefit for the United States in terms of adaptation needs.

In specific cases, limiting and adaptation strategies can offer symbiotic benefits, particularly in the energy sector. For instance, distributed energy-generation systems can enhance energy efficiency by allowing waste heat from power production to be used for water heating and other purposes. At the same time, these distributed systems strengthen resilience against climate change impacts by reducing the risk of widespread power loss from severe storms or from peak periods of demand during heat waves. Similarly, advances in the energy efficiency of cooling systems can both constrain the growth of GHG emissions and help to affordably meet the greater need for air-conditioning that may result from global warming.

There can also, however, be trade-offs between mitigation and adaptation strategies. For instance, in the example above, strategies to promote more widespread use of air-conditioning (to adapt to higher summer temperatures) would undermine mitigation goals if the added energy demand is met by GHG-intensive power sources. Hamin and Gurran (2009) examined existing land use–related strategies that cities across the United States and Australia have taken or proposed in response to climate change, and they found many examples of actions that led to conflicts between mitigation and adaptation goals.

International Ancillary Benefits and Costs

Achieving significant reductions in atmospheric GHG concentrations requires action by many countries around the world. As is the case in the United States, actions to reduce GHG emissions in these other countries could create ancillary benefits or costs for energy, air pollution, agriculture and forestry, and adaptation. In a few such cases, the United States may directly benefit from the actions of other countries. Both the Third and Fourth Assessment Reports of the Intergovernmental Panel on Climate Change (IPCC, 2001, 2007a) survey the value of ancillary benefits attributable to actions designed to limit climate change. Those reviews point out that the literature on this subject is limited and the studies that do exist vary widely in modeling approaches, key assumptions, and coverage. As a result, quantifying benefits, especially for middle- and low-income countries, is very difficult.[4] Nevertheless, some qualitative conclusions are possible. In some cases, a country's actions to reduce GHG emissions produce direct, country-specific benefits. For example:

- In middle- and low-income countries in particular, the monetized health benefits of reducing emissions often nearly offset the costs of GHG reduction. The health benefits are especially great in situations where emissions reduction has a strong impact on population exposure to pollutants such as black carbon, for instance from domestic stove heating and cooking. The IPCC suggests that the health benefits from such actions are typically 40 times greater than health benefits from reducing emissions from central, tall-stack power facilities.

- Reducing fossil fuel combustion lowers not just CO_2 emissions but also emissions of the pollutants that form tropospheric ozone, and this can have benefits for human health and for agriculture. The IPCC indicates that the agricultural benefit of reducing ozone pollution could substantially offset the welfare loss that poor, rural households may experience from the costs of actions to limit GHG emissions.

- GHG emissions-limiting strategies could significantly reduce the cost of controlling conventional air pollution emissions if controls on both types of emissions are integrated into one system. This benefit would be greatest in regions where conventional pollutants are not yet controlled (largely in low-income countries).

---

[4] In particular, Section 8.2.4 of the Working Group III report, the IPCC Third Assessment Report, contains a critical review of the problems involved in estimating ancillary benefits. More recently, the International Energy Agency (IEA) World Energy Outlook contains data on the estimated deaths from indoor cooking and on the environmental effects of fuelwood use. The IEA data are available at *http://www.worldenergyoutlook. org/implication.asp.*

- Reducing rates of deforestation internationally may significantly decrease runoff and land degradation, as well as enhancing species diversity. Reducing tillage intensity can also have water quality and soil conservation benefits (although these benefits are very difficult to quantify in monetary terms).

The EPA has an Integrated Environmental Strategies program that seeks to develop country-specific strategies for capturing the ancillary benefits of limiting both climate change and air pollution. The results of studies done in eight countries are consistent with the data developed by the IPCC.[5]

In other cases, the ancillary benefits of GHG limitation have spillovers that result in the benefit being shared among nations. For example:

- To the extent that any country reduces oil consumption, all countries may benefit from a downward pressure on world oil prices. However, the magnitude of this benefit for any particular country depends upon the oil dependence of the country's economy, as well as independent forces such as price shocks arising from supply disruptions, and the possibility that lower oil prices may stimulate increased demand.
- Methane is not only a powerful GHG but also a precursor to tropospheric ozone (which, as noted earlier, adversely affects human health and agriculture). Reducing methane emissions thus can both improve air quality and limit overall GHG concentrations. Barker and Bashmakov (2007) suggest that the health benefits may exceed the marginal cost of methane reductions.
- Hydrofluorocarbons (HFCs) are increasingly being used to replace refrigerant chemicals that are banned due to their impacts on the stratospheric ozone layer. However, HFCs are also strong greenhouse gases. Velders et al. (2009) estimates that global HFC emissions in 2050 could be as high as 5.5 to 8.8 Gt CO_2-eq per year. Most of the future growth in HFC emissions is expected to take place in low-income countries. Developing alternatives to HFCs in these countries could therefore make an important contribution to limiting GHG concentrations.

There are a number of other possible ancillary costs and benefits of limiting GHG emissions, but their value is even less certain than those listed above. On balance, however, there appear to be significant ancillary benefits from reductions in GHG emissions internationally. Many of these benefits are largest in middle- and low-in-

---

[5] See *http://www.epa.gov/ies/pdf/general/IES%20Bangkok%20April%2023%20final.pdf* for a summary of the IES work.

come countries and are directly related to a country's actions to reduce its own GHG emissions.

KEY CONCLUSIONS AND RECOMMENDATIONS

Policies to limit the magnitude of climate change may offer direct ancillary benefits such as reducing the emission of air pollutants and lowering dependence on imported petroleum fuels. The use of offsets as a climate policy may have indirect but beneficial effects on forestry and agricultural practices. In principle, climate change limiting policies should be designed to capitalize on these benefits, but applying this principle systematically is difficult in practice, because estimates of ancillary benefits are uncertain and the benefits, costs, and potential for leakage are often project- and location-specific. Nevertheless, ancillary benefits may be robust enough to justify influencing national climate policy in a few areas, including the following:

- *To accelerate the reduction of oil use in the transportation sector.* The combined costs of oil consumption impacts on U.S. energy security and human health (as estimated by recent studies discussed in this section) is roughly equivalent to 45-55 cents per gallon of gasoline (which can be converted to roughly $45-55 per ton of CO_2). Policy strategies for actually realizing energy security and health benefits from reducing oil use requires more thorough analysis than is presented here, but the possible benefit is large enough to warrant attention.
- *To capture the full benefit of emissions-limiting actions, especially in low- and middle-income countries.* Although there is considerable variation among countries, ancillary benefits associated with reducing air pollution and improving forestry and agricultural practices can be particularly large in low- and middle-income countries.
- *To reduce emissions of methane, short-lived pollutants, and HFCs.* Methane, other precursors for tropospheric ozone, and black carbon aerosols lead to adverse human health effects on broad regional scales. HFCs, used as replacements for stratospheric ozone-depleting refrigerant agents, are potent GHGs.
- To encourage climate policy actions that result in closure or upgrading of electric power plants with disproportionately large health impacts. This applies primarily to power plants that lack effective air pollution controls and are located near densely populated areas.

These potential co-benefits provide additional impetus for pursuing several of the key actions suggested elsewhere in this report, including, for instance, the reduction of oil use in the transportation sector; the strategies for engaging middle- and low-income countries in international climate change agreements; the development of global agreements for reducing emissions of species such as methane, tropospheric ozone precursors, black carbon, and HFCs; and the efforts to retire existing carbon-intensive infrastructure (in particular, poorly controlled power plants located in densely populated areas).

EQUITY AND EMPLOYMENT IMPACTS

When considered in a long-term global context, climate change presents an array of challenging ethical dilemmas; for example, debates about how to fairly allocate responsibilities for reducing GHG emissions among low- and high-income countries have stymied international negotiations for years. Likewise, many argue that climate change is, at its core, a question of intergenerational equity: To what degree should current generations take action to protect future generations from harm?

While these are tremendously important issues to grapple with, this section focuses more narrowly on a set of concerns of particular interest to U.S. policy makers—how policies for reducing domestic GHG emissions may alleviate or exacerbate equity and "environmental justice" among different parts of American society. We examine how such policies will cause different impacts across regions and population groups due to variations in the nature and carbon intensity of regional economic activity, and in the resources and adaptability of different populations. We then focus on examining how policies to limit climate change may affect economic and employment opportunities across the country, since employment opportunities are of course a key means of enhancing equity.

Socioeconomic Distributional Impacts

Climate change impacts, and actions to limit these impacts in the future, will take place in the context of existing social and economic disparities, many of which are related to environmental concerns. In the United States, households in inner cities and rural areas, and African- and Hispanic-American households, are disproportionately poor. In metropolitan areas, poor households locate where housing costs are lowest, often in zones of heavy industry or near noxious facilities (e.g., waste treatment plants, transportation facilities, and power plants); as a result, exposure to air pollution and toxics is greater among low-income and minority households (Brulle and Pellow, 2006; Schweitzer and Stephenson, 2007). Poor households in rural areas suffer severe mobility problems due to limited access to private vehicles or public transit services, which in turn restricts access to jobs, health care, education, and other basic services.

Lower-income households, on a per capita basis, consume less energy and hence contribute less to GHG emissions. Yet low-income and minority populations are likely to suffer disproportionately from climate change effects. Some examples include the following:

- *Extreme heat or cold events.* Studies of extreme heat events show elevated mortality and morbidity risk for small children, the elderly, and African Americans (Basu and Ostro, 2008; Kovats and Hajat, 2008) due largely to heat island effects in urban areas, heat exposure in outdoor work, and less access to air-conditioning. As extreme heat and cold events increase, the burden of paying for air-conditioning or home heating will increase, and the poorest households will be least able to absorb these additional costs.
- *Air pollution.* Because some air pollution is a function of weather conditions, related health impacts may increase under climate change. Poor and minorities will be disproportionately affected by such changes, due to greater pollution exposure levels and limited access to health care.
- *Low-skill/low-wage jobs.* Climate change may impact certain jobs disproportionately through effects on agriculture, tourism, and other sectors that use low-skill/low-wage labor; for example, rising sea level will affect coastal zones, declining precipitation may affect winter recreation areas, and drought conditions will affect agricultural activity, all with resulting unemployment risks.
- *Food, water, and energy.* Both climate change itself, and policies to limit climate change, may lead to rising prices for water, food, and energy. Since poor households spend a greater proportion of income on these essentials, they will suffer a disproportionate impact of such price increases (Hoerner and Robinson, 2008; Morello-Frosh et al., 2009).
- *Disasters.* Poor and minority households are often more vulnerable to weather-related disasters. For instance, in the case of hurricanes, those who do not speak English, do not have geographically dispersed social networks, and do not have personal vehicles are less likely to evacuate. Poor households are less likely to have insurance to cover disaster losses (Elliott and Pais, 2007).

Because our understanding of the equity outcomes of GHG emissions-limiting policies is currently quite limited, it is instructive to examine studies of how other sorts of environmental protection regulations have affected equity outcomes. Some examples, focused primarily on analyses of local impacts of air quality improvements on different social groups in Southern California, include the following:

- Pearce (2003) lists a number of studies that consider whether the results of command and control environmental policies can be considered "elitist goods" that benefit the rich more than the poor. He concludes that environmental and safety measures tend to benefit households in lower income brackets more than households with higher incomes.
- Sieg et al. (2004) measured the response to the reduction of O_3 concentration following the implementation of the 1990 Clean Air Act Amendments (CAAA). He concludes that household wealth plays a relevant role in the distribution of environmental improvement-related welfare; for instance, low-income renters benefited less from O_3 reductions, while low- and high-income homeowners gained significantly due to the appreciation of their property.
- Tran (2006) found that the distribution of relative welfare gains from the 1990 CAAA is fairly even across income groups.
- Fowlie et al. (2009) examined the correlation between NO_x emissions reductions and social structure of the areas where they occurred and concluded that race and income are not correlated to significant differences in emissions reduction, similar to Tran's assessment.
- Shadbegian et al. (2005) studied costs and benefits of the SO_2 trading scheme among coal-fired power plants in the Midwest using data on abatement costs and estimating cancer risk reduction benefits. He estimated that the cost-benefit ratio for minorities and the elderly is not substantially different from that of other social groups, but the cost-benefit ratio for the poor is slightly less favorable than for the average individual.

Such studies thus yield conflicting results and no clear evidence of systematic biases in adverse impacts upon low-income and minority populations. As discussed below, however, the unique characteristics of climate change policy require closer investigation into potential distributional impacts. The distributional impacts of carbon-pricing policies depend on the structure of the program. In the case of cap and trade, impacts would depend on the scope of the program across industry sectors, how initial allowances are distributed, the stringency of the cap (which affects downstream prices to consumers), and how revenues are spent. As discussed in Chapter 4, initial allowances may be allocated via auction or be given away. Pricing the initial allowances provides revenues to government, and these revenues could be spent in a variety of ways, from providing energy allowances to low-income households to reducing income or corporate taxes (Burtraw et al., 2008). Low-income households spend a larger proportion of income on energy-related consumption than middle- or high-income households. Shammin and Bullard (2009) estimate that direct energy use accounts for 12 percent of household income for those in the lowest income quintile and accounts for 4 percent of household income for those in the highest quintile. Thus, any policy that increases the price of energy will be regressive, absent some form of revenue redistribution.

Because national carbon-pricing schemes do not yet exist, studies of their impacts must rely on modeling exercises. Most such modeling studies take into account direct and indirect costs of carbon allowances or equivalent carbon taxes (e.g., increases in the price of energy and the prices of goods and services that require energy), but they do not attempt to measure the indirect benefits of the carbon price, such as the benefits of reduced air pollution, nor do they analyze effects on capital and labor or estimate possible substitutions or changes in purchasing behaviors.

Studies that use household consumption data and similar methods and assumptions show similar results (Burtraw et al., 2008; Chamberlain, 2009; Dinan and Rogers 2002; Grainger and Kolstad, 2009; Shammin and Bullard, 2009) (see Table 6.1 for summary). A carbon price is consistently regressive, because lower-income households use a larger proportion of their earnings to purchase energy-intensive products (gas and electricity being the most important). The extent of regressivity depends on whether initial allowances are given away and how tax revenues are spent. Grainger and Kolstad (2009) add that, if the effects of carbon trading are estimated on a per capita basis, the regressive effects are even more relevant. Chamberlain (2009) claims that the cap-and-trade burden will be heavier on younger households, on single parents, and on families living in the Northeast and in the South.

Burtraw et al. (2008) found that all scenarios are mildly regressive before revenue redistribution, but the net effects vary. Free allocation to emitters and using revenues to reduce income taxes are the most regressive alternatives. In the case of free allocation, benefits accrue to stockholders who are disproportionately higher income. Reduced income taxes work similarly, because the highest income brackets receive the greatest tax savings. In contrast, if revenues are used to expand the earned income tax credit, the result is progressive. Rose and Oladosu (2002) generate similar results, with free allocation of allowances leading to the most regressive outcomes. Their results show that lower-income families will suffer from increased energy and household prices, but middle-class households will bear a higher relative burden. Additionally they argue that, if other emissions-limiting measures and carbon sequestration are taken into account, the impact of energy prices on incomes would be less relevant.

Most analysts (Boyce and Riddle, 2007; Burtraw et al., 2008; Dinan, 2009; Grainger and Kolstad, 2009; Shammin and Bullard, 2009) agree that, if allowances in a cap-and-trade scheme are auctioned, the revenues can be used to offset the regressivity of the policy. A variety of policy instruments would be necessary to target most low-income households, taking into account the fact that some of them earn income (and so could be compensated through income taxes rebates or other income-related forms), and others do not (and could be compensated by adding to the existing support programs). The choice of instruments should be oriented to limit administrative and compliance costs (Dinan, 2009).

The studies above—all looking at gross income categories and large spatial units—suggest that, with appropriate policy instruments, the regressivity of a carbon price can be offset, at least on average. But these results do not necessarily tell the whole story of how particular vulnerable communities or population groups may be affected. A concern, for instance, raised by leaders in the environmental justice movement is that a carbon-pricing system may do little to alleviate the disproportionate burden of pollution emissions currently being faced by some low-income and/or minority communities (Morello-Frosch et al., 2009).

When viewed at a global scale, it makes little difference where CO_2 emissions are reduced—the CO_2 emitted from any point source mixes into the atmosphere and does not directly increase risks on a local scale. But how and where emissions occur have important consequences for the communities where the associated copollutants (e.g., particulates, air toxics) accrue and can lead to localized "hot-spot" health impacts. For instance, coal-fired power plants, which release more than 40 percent of total U.S. CO_2 emissions, also emit large concentrations of sulfur and nitrogen oxides, mercury, and other air toxics that are known or suspected carcinogens and neurotoxins. A growing body of research has demonstrated that such risks are not evenly distributed across the population. Rather, people of color and low-income communities often face higher-than-average health risks from local pollution sources (Bullard, 2000; Pastor, 2007)(see previous section for discussion of the air quality co-benefits stemming from CO_2 emissions-reduction efforts).

Because climate change may increase urban air pollution problems and further exacerbate this unequal health burden, some suggest that GHG emissions-reduction efforts should be designed to ensure that advanced, cleaner technologies (for power plants, factories, etc.) are directed to the most polluted neighborhoods (e.g., Morello-Frosch et al., 2009). The Clean Air Act and some state-level air quality management programs do address pollution hot-spot concerns with requirements for more stringent pollution abatement actions; however, there may be opportunities for accelerating such efforts via careful coordination with programs targeted at GHG emissions abatement.

TABLE 6.1 Summary of Impact Analyses of Cap and Trade and Carbon Taxes in the United States

Author	Data/Method
Dinan and Rogers (2002) Cap and trade Free distribution of C allowances; the government captures 45% of the profit through corporate income taxes	Microdata on annual expenditure and consumption of households (Consumer Expenditure Survey [CES]). Input-output model (Bureau of Economic Analysis [BEA]). Base year 1998.
Rose and Oladosu (2002) Cap and trade Free distribution of C allowances	Data on emissions and energy from Energy Information Administration (EIA) and EPA. Data on income from Internal Reveune Service. Computable general equilibrium model with 41 sectors and 10 income brackets.
Grainger and Kolstad (2009) Carbon tax	Microdata on annual expenditures and consumption of households (CES: 2003). Emissions factors and 1997 data on the structure of the U.S. economy to calculate how a price on carbon is ultimately distributed across income groups. Input-output table (BEA).

Target	Assumption	Results
Reduce CO_2 emissions by 15% below 1998 levels.	100% allowance costs passed on to the consumer only for the private sector. Labor and capital constant, purchasing behavior unchanged.	Only domestic C trading: Price of one allowance = $100 Increase of household costs due to allowances cost as % of income: 1st quintile 6.6% $ 558 2nd quintile 3.7% $ 719 3rd quintile 3.1% $ 955 4th quintile 2.7% $1,236 5th quintile 1.7% $1,802 Average: 2.9%
Reduce CO_2 emissions by 10% below 1990 levels.	Substitutions and changes in behaviors are taken into account.	Price of one allowance = $128 per ton of C. Lower-income families fare worst than high-income families, but middle-class households suffer the major relative impact.
Incidence of a $15/ton CO_2 tax		$15 per ton of CO_2, equivalent to approximately $55 per ton of C. The poorest quintile's burden as share of annual income is 3.2 times that of the wealthiest quintile. An average household in the lowest income quintile would pay around $325 per year, while an average household in the wealthiest quintile would pay $1,140 annually. On an annual basis, a carbon price is 2-3 times more regressive than on a lifetime basis (i.e., using annual expenditures)

Continued

TABLE 6.1 Continued

Author	Data/Method
Burtraw et al. (2008) Cap and trade	Microdata on annual expenditures and consumption of households (CES: 2004-2006). Direct costs based on EIA energy prices and estimated cost of carbon considering elasticities. Indirect costs based on emission intensity from Hassel (2009).
Chamberlain (2009) Cap and trade	Consumer expenditure data for calendar year 2006. 2002 benchmark input-output accounts from the BEA, released in September 2007, inflated at 2006.
Metcalf (2009) Carbon tax	Consumer expenditures and input-output table.
Shammin and Bullard (2009) Cap and trade	Microdata on annual expenditures and consumption of households (CES: 2003). Sample of the families that reported four quarters in a row. Emissions factors and 1997 data on the structure of the U.S. economy according to input-output table (BEA).

Target	Assumption	Results
Incidence of climate policy in the year 2015 with an hypothetical emissions cap in the figure is set at about 75 percent of baseline emissions.	100% allowance costs passed on to the consumer in every sector but electricity. In electricity, 80% allowance cost passed on to the consumer. Labor and capital constant. Changes in consumer expenditures estimated, new Corporate Average Fuel Economy standards included as change in transportation costs. Consumer surplus lost estimated.	Suits Index (when negative means regressivity) for the CO_2 price of $20.87 is -0.18. Low-income families in the Northeast and mountains will experience higher losses in percentage on their income.
Reduce C emissions by 15 percent compared to 2006 levels.	100% allowance costs passed on to the consumer. Labor and capital constant, purchasing behavior unchanged.	Allowance price of roughly $100 per metric ton of C that corresponds to $27.27 per ton of CO_2. 1st quintile 6.2% $ 617 2nd quintile 3.2% $ 863 3rd quintile 2.0% $ 1,100 4th quintile 2.0% $ 1,418 5th quintile 1.4% $ 2,091
Incidence of a $15 per ton CO_2 tax	100% allowance costs passed on to the consumer. Labor and capital constant, purchasing behavior unchanged.	The bottom half of the population faces losses in after-tax income ranging from 1.8% to 3.4% of its income. Top half of the population faces losses between 0.8% and 1.5% of its income. Based on average income, a carbon tax does not appear to disproportionately burden one region of the country.
Incidence of a $ 100/ton C allowance	100% allowance costs passed on to the consumer. Labor and capital constant, purchasing behavior unchanged.	At $100/ton C allowance price, a household with an income of $36K would face a cost-of-living increase of $915: 1st quintile 3.4% $ 465 2nd quintile 2.8% $ 700 3rd quintile 2.5% $ 915 4th quintile 3.1% $ 1,607 5th quintile 1.9% $ 1,905

Region-Based Distributional Impacts

Carbon pricing will also have impacts that vary across regions, due to differences in industry mix, energy sources, and climate characteristics. For example, regions dependent upon coal-based energy will be more adversely affected than regions with more diversified energy resources. Conversely, regions that produce "green" energy (wind, switchgrass) will benefit, particularly if tax revenues are used to subsidize these energy sources. Regional impacts will also depend on climate, with mild climate regions (such as much of the West Coast) less affected due to lower household energy demands. Low-density development patterns typical of many newer cities are associated with more private vehicle travel and, thus, more vulnerability to rising gasoline prices.

The combination of low-density development and heavy reliance on coal-generated electricity results in large metropolitan carbon footprints in particular regions, such as the Ohio Valley and the South (Brown et al., 2009b). For example, the average resident in Lexington, Kentucky, emits 2.5 times more carbon from transport and residences than the average resident in Honolulu, Hawaii. When adjusting for a metropolitan area's economic output, there is up to a fourfold variation among these urban carbon footprints. Thus, implementation of a price on carbon will have highly variable impacts across regions.

Burtraw et al. (2008) estimated the average social welfare loss (measured as percent of average income) across 11 U.S. regions for four different policy scenarios (Table 6.2). They found that impacts do vary across regions and across scenarios. In general, welfare losses are smallest in regions with lower rates of energy consumption (California/ Nevada and the Northwest) and greatest in regions with higher rates of energy consumption and more dependence on coal-based energy (Ohio Valley, Plains).

A second example is provided by the EPA's analysis of the 2008 Lieberman-Warner Climate Security Act (EPA, 2009), which is comparable to the 200 Gt CO_2-eq scenario discussed in Chapter 2 (with a carbon price of approximately $60 per ton). An economic model (ADAGE) was used to estimate regional gross domestic product (GDP) and consumption changes across five U.S. regions for target years 2020 and 2040 (Figure 6.1). The results indicate that climate change limiting policies will have varying impacts across different states and regions. In this scenario, the Plains region (North Dakota, south through Texas, plus Minnesota) suffers the greatest losses. Differences across regions are attributed to variations in industry mix, energy consumption, energy sources, and assumptions regarding allocation of allowances.

Employment Impacts

An equity concern of climate policies is the question of potential effects on employ-ment. Some groups (e.g., National Association of Manufacturers) claim that national climate change response policy measures will lead to job losses numbering in the millions. At the same time, "green jobs" advocates argue that responding to climate change provides an unprecedented opportunity to create new industries and new jobs across the country, for example, in manufacturing and installing solar and wind power systems, weatherizing homes and commercial buildings, and building and operating public transportation systems. It is argued that a major expansion of these sorts of green jobs will help create the national workforce base that our country needs to make a major transition to a low-carbon economy (see Chapter 5) and to become a world leader in the development, use, and export of clean energy technologies (Gold et al., 2009). It is likewise suggested that investments in energy efficiency and renew-able energy technologies (per dollar invested) generate more domestic employment than investments in fossil fuel production industries and can create more opportuni-ties for "pathways out of poverty" for low-income Americans (Pollin et al., 2008).

The Energy Modeling Forum (EMF22) modeling study of the macroeconomic impacts of climate change limiting policies (presented in Chapter 2) indicates little impact on U.S. GDP through 2050. These results imply that net employment impacts will thus also be small, most likely because job creation in "green" industries will be offset by losses in other sectors. Small net effects, however, mask the significant shifts in em-ployment opportunities and challenges that some economic sectors and geographic regions will face.

Figure 6.2 shows that, for a climate change limiting scenario of 203 Gt CO_2-eq, the major negative impacts are in fossil fuel–related industries and in energy-intensive in-dustries such as utilities and agriculture. Note that the total estimated loss is less than $1 trillion, compared to a national output in 2030 of about $36 trillion (but also note this scenario allows for unlimited domestic offsets, which lowers the carbon price). The largest losses are in the fossil fuel industries, agriculture, machine manufacturing, and wholesale and retail trade. These sectors, and the regions where these sectors are more concentrated, would be expected to suffer the largest job losses. Given the con-centration in relatively few sectors, it follows that employment impacts will be highly uneven.

The magnitude of energy system changes required to achieve GHG reduction goals suggests the need for significant restructuring of the economy. Previous major restruc-tures, such as the deindustrialization of the 1970s and 1980s, provide some guidance

TABLE 6.2 Average Social Welfare Loss by Region (as percent of average income for four scenarios)

Region	Scenario			
	(i) Reduce Income Tax	(ii) Earned Income Tax Credit	(iii) Exempt Transport Sector	(iv) Free Credit Allocation
Southwest	1.40%	1.27%	1.32%	1.44%
California/Nevada	0.97%	1.23%	1.25%	1.25%
Texas	1.59%	1.27%	1.39%	2.00%
Florida	**1.68%**	1.56%	1.59%	1.86%
Ohio Valley	1.65%	**1.78%**	**1.79%**	1.89%
Mid-Atlantic	1.03%	1.45%	1.48%	1.41%
Northeast	1.09%	1.68%	1.66%	1.51%
Northwest	0.90%	1.01%	1.05%	1.14%
New York	0.95%	1.27%	1.38%	1.29%
Plains	1.42%	1.72%	1.71%	1.59%
Mountains	1.53%	1.38%	1.54%	1.59%
National	1.36%	1.36%	1.43%	1.60%

NOTES: Scenarios include the following: (i) reduce income tax by the amount of revenue generated from the cap-and-trade program, (ii) expand the earned income tax credit by the amount of revenue generated from the cap-and-trade program, (iii) exempt the transport sector from the cap-and-trade program, and (iv) allocate all initial credits for free to corporations midstream or upstream in the fuel cycle. In all scenarios it is assumed that the federal government retains 35 percent of the allowance value. Policies begin implementation in 2008, and the scenario target year is 2015. The resulting carbon price per ton is about $41 (in 2006 dollars), and the reference basis for the calculations is a scenario with no carbon pricing policy. Green indicates the regions experiencing the most positive impacts, while red indicates the regions experiencing the most negative impacts.
SOURCE: Adapted from Burtraw et al. (2008).

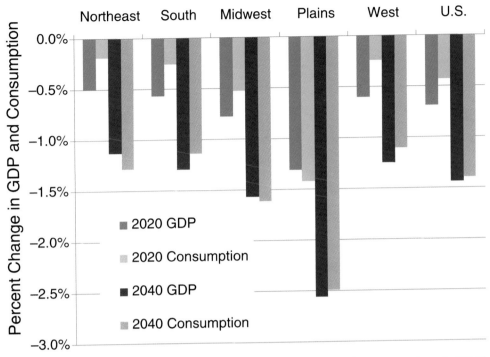

FIGURE 6.1 Estimated changes in GDP and consumption by U.S. region, for an approximately 200 Gt CO_2-eq budget. Differences across regions are attributed to variations in industry mix, energy consumption, energy sources, and assumptions regarding allocation of allowances; the largest losses are projected for the Plains states. SOURCE: Adapted from the EPA analysis of the 2008 Lieberman-Warner bill (S.2191). Available at *http://www.epa.gov/climatechange/downloads/s2191_EPA_Analysis.pdf.*

on the types of impacts on jobs and workers that might occur. Decline of the manufacturing sector resulted in the loss of relatively higher-wage skilled blue-collar jobs, many of them concentrated in the industrial belt of the U.S. Northeast. Studies have identified a shift in the wage distribution as relative demand for skilled blue-collar workers has declined, while growth of the service sector has generated demand for both highly skilled technicians and low-skill laborers (Appelbaum and Alpin, 1990; Noyelle, 1987).

Impacts on employment will depend on the capacity of the workforce to adjust to rapidly changing circumstances. Those whose jobs are eliminated may or may not have the appropriate skills for emerging jobs, and these jobs may emerge in other regions. In the 1970s and 1980s, job losses were concentrated in the industrial Northeast, while job growth took place in the South and West. Thus, much of the adjustment involved migration of populations. Younger, more educated and skilled workers are more likely

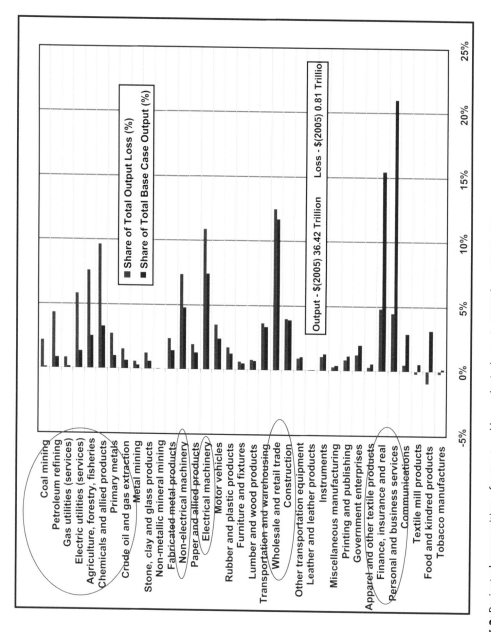

FIGURE 6.2 Projected composition of output and losses by industry in 2030, for a 203 Gt CO_2-eq scenario with unlimited domestic offsets. These projections indicate that fossil fuel industries, agriculture, machine manufacturing, and wholesale and retail trade sectors—and the regions where these sectors are more concentrated—would be expected to suffer the largest job losses. SOURCE: Goettle and Fawcett (2009).

to migrate, leaving an older, less educated and less skilled population behind. One social equity concern is the possibility that already distressed cities could be further affected by job losses from policies aimed at limiting climate change.

A recent Congressional Budget Office report that discusses potential impacts of climate limiting policies notes that adjustments associated with the decline of manufacturing jobs took years to accomplish and had significant adverse effects on some workers (CBO, 2009). It seems likely that similar impacts may occur as a result of economic restructuring associated with climate change limiting policies, and these may be concentrated in specific regions and occupations.

At the same time, it seems inevitable that new jobs will be created through investment in improving the energy efficiency of industry, buildings, and transportation and in construction of new power plants and energy infrastructure. Making robust predictions about future job generation, however, is a difficult task. First, it is complex even to define "green jobs," a concept that could encompass economic activities as diverse as insulating buildings, synchronizing traffic signals, conducting research on carbon capture and storage, or manufacturing more energy-efficient plasma screen televisions. More generally, it is difficult to forecast factors such as the employment effects of investments in new technologies, the specific mix of new jobs across occupations or skill levels, or the particular characteristics of the new green economy (for instance, we cannot foresee future technology breakthroughs, or how carbon prices will affect production or consumption patterns).

A recent study by the Pew Center on Global Climate Change (Pew Center, 2009c) tabulated growth of the "Clean Energy Economy" under five categories: clean energy, energy efficiency, environmentally friendly production, conservation and pollution mitigation, and training and support. As of 2007, some 68,200 businesses across all 50 states accounted for about 770,000 jobs (private-sector employment in 2007 was about 114 million [U.S. Bureau of Labor]). From 1998 to 2007, job growth was 23 percent for clean energy, 18 percent for energy efficiency, and 67 percent for environmentally friendly production. The Pew study confirms that the clean energy economy can in fact be an engine for new economic growth, and these studies are largely consistent with expectations of economic restructuring described above.

Given that new industries and jobs will develop, it is appropriate to consider whether this investment can be used to promote local economic development in high-poverty areas such as inner cities, Appalachia, or Native American reservations, or in the areas that are expected to experience the greatest economic and job losses as a result of policies to limit climate change. Absent deliberate government policy, we would expect new industries to locate competitively (e.g., based on labor force quality and access, land prices, taxes, access to transport networks, etc.). High-poverty areas are typically not competitive. Inner cities suffer from a lack of a skilled labor force, high land prices and local taxes, and often poor public services. Rural areas have limited labor force and transport network access.

The historical record on economic development efforts in these disadvantaged areas shows that significant, targeted, coordinated, and long-term investments in education, training, and capacity building would be required to be successful. Economic development efforts have taken many forms. Local governments may offer financial assistance or other incentives to businesses (e.g., tax abatement, loans, and grants). In some cases cities invest in major venues (e.g., sports stadia and convention centers) to promote local development. Enterprise zones have been established in economically distressed cities, typically offering financial incentives, special permitting or zoning, infrastructure, and tax credits for job generation or net revenue generation (e.g., Bartik, 1991; Wassmer, 1994; Wren, 1987). The outcomes of economic development programs have exhibited mixed degrees of success. The key explanation for their lack of greater success is that most efforts are driven by public policy goals rather than an understanding of local markets and economies (e.g., Dewar, 1998; Peters and Fisher, 2002; Wolman and Spitzley, 1996).

There are many anecdotal examples of unique programs around the United States aimed at fostering employment in at-risk communities by providing training and placement in green jobs (see Box 6.1). These programs merit careful evaluation to gauge their long-term success and the potential for more widespread implementation.

The Need for Broad Political Participation

As illustrated in the preceding sections, policies that are sufficiently stringent to meet GHG emissions-reduction goals will impose new costs and benefits across industries, regions, and population groups. Stakeholders representing these constituencies can be expected to be actively involved in the policy process. The history of environmental policy is illustrative. The U.S. automobile industry, for example, initially fought against fuel efficiency standards but, once it became evident that standards would be implemented, used their political influence to affect targets and implementation dates (Howitt and Altshuler, 1999). The "auto lobby" (auto manufacturers, suppliers, oil companies, and highway construction) has remained a powerful influence on emissions, safety regulation, and fuel tax policy to the present day (Sperling and Gordon, 2009).

Not all stakeholders, however, are influential or even present in the policy-making

---

BOX 6.1
Examples of Programs to Create Employment in Green Industries

Sustainable South Bronx and the BEST Program, South Bronx, New York. The Bronx Environmental Stewardship Training (BEST) Program with Sustainable South Bronx is one of the nation's first green-collar job training and placement systems. Students graduate with certifications such as water quality management and Occupational Safety & Health Administration Brownfield Remediation. The program is aimed at bringing those with little or no work experience, or those with prison records, into the workforce.

Solar Richmond, Richmond, California. Solar Richmond was founded to provide green-collar job training and placement in solar photovoltaics and solar thermal installation jobs. Their intensive training course includes hands-on work on real solar installations. They specifically target installations for low-income residents, churches, and schools, to help these groups reap the rewards of lower energy costs and local job creation. Graduates of the program are given the opportunity to work for Solar Richmond.

St. Patrick Center, St Louis, Missouri. The St. Patrick Center, the largest homeless service provider in Missouri, is launching Project GO!Green, in which homeless individuals are trained in horticultural infrastructure and urban farming. The program is now expanding to serve professionals who have recently lost their jobs.

Milwaukee Area Investment Board & Student Conservation Association, Milwaukee, Wisconsin. This is a green job training program for high school–aged youth. The work mainly focuses on conservation projects. Summer employment is salaried for all participants.

OAI YouthBuild and the City of Chicago's Green Corps, Chicago, Illinois. Brownfield remediation and environmental certifications are all provided by Chicago's OAI, inc. A grant from the U.S. Department of Labor was recently awarded to the YouthBuild Program, which targets at-risk youth with a comprehensive training and job placement system. This program directly links into the City of Chicago's Green Corps that is championed by Mayor Daley.

---

process. The poor are less likely to vote, to be aware of relevant policy decisions, to understand the policy process enough to effectively participate, or to have the resources (i.e., time, information, and money) necessary for political participation (e.g., Bartels, 2008). The environmental justice problems of toxics and hazardous facilities location discussed above are attributed at least in part to the lack of political power of the affected communities to influence environmental monitoring or facility location decisions (Capek, 1993; Zimmerman, 1993). Without a "seat at the table," the needs and concerns of these groups are easily ignored.

One specific example to consider: Pursuing the types of climate change limiting strategies discussed in this report will require siting and building a variety of new facilities—power plants, carbon sequestration facilities, production sites for renewables, and expansion of the electric grid—all of which are subject to environmental review. Because environmental review for major public projects can take years or even decades to complete (Altshuler and Luberoff, 2003), and lengthy review processes add to costs and project risk, there is growing discussion of streamlining these processes. Environmental justice advocates have raised concerns that this sort of streamlining can lead to less powerful interest groups being more easily ignored.

Given the numerous ways in which currently disadvantaged groups could be adversely affected by policies for limiting climate change (as well as by the impacts of climate change itself), careful attention needs to be paid to procedural equity concerns and efforts to ensure full engagement of disadvantaged populations. The affected parties need a seat at the table in discussions of how to avoid harmful impacts from the outset, or how to correct for unanticipated adverse impacts that may arise. Ensuring access of low-income and other disenfranchised populations to programs and incentives for reducing energy demand and utilizing low-carbon energy technologies is not just a matter of fairness. It is also a matter of practical necessity, as achieving major GHG emissions reductions will be very difficult unless all segments of American society are participating in these efforts.

Equity concerns raise a number of substantive policy design and implementation issues to be considered by policy makers. This includes, for example, consideration of policies that redistribute revenues to low-income households to offset the regressive effects of higher energy prices, policies for avoiding co-pollutant hot spots, policies that create new clean-energy jobs and industries in disadvantaged communities, and policies to avoid further penalizing the already limited mobility of many poor and minority communities. Processes for establishing GHG emissions-reduction policies should thus include broad, sustained public participation efforts. There is a substantial literature about the mechanisms for effective public participation in environmental decision making (e.g., Beierle, 1998; NRC, 2008) to which we refer the reader for further consideration of this issue.

KEY CONCLUSIONS AND RECOMMENDATIONS

Low-income groups consume less energy per capita and therefore contribute less to associated GHG emissions. Yet, low-income and some disadvantaged minority groups are likely to suffer disproportionately from adverse impacts of climate change and, absent proactive policies, may also be adversely affected by policies to limit climate change. For instance, energy-related goods make up a larger share of expenditures in poor households, so raising the price of energy for consumers may impose the greatest burden on these households. Likewise, limited discretionary income may preclude these households from participating in many energy-efficiency incentives. Because these impacts are likely but not well understood, it will be important to monitor the impacts of climate change limiting policies on poor and disadvantaged communities and to adapt policies in response to unforeseen adverse impacts. Some key strategies to consider include the following:

- Structuring policies to offset adverse impacts to low-income and other disadvantaged households (for instance, structuring carbon-pricing policies to provide relief from higher energy prices to low-income households); designing incentive-based climate change limiting policies to be accessible to poor households (such as graduated subsidies for home heating or insulation improvements);
- Assuring that efforts to reduce energy consumption in the transport sector avoid disadvantaging those with already limited mobility; and
- Actively and consistently engaging representatives of poor and minority communities in policy planning efforts.

Major changes to our nation's energy system will inevitably result in shifting employment opportunities, with job gains in some sectors and regions but losses in others (i.e., energy-intensive industries and regions most dependent on fossil fuel production). Policy makers could help smooth this transition for the populations that are most vulnerable to job losses through additional, targeted support for educational, training, and retraining programs.

CHAPTER SEVEN

Multilevel Response Strategies

The primary focus of this report is on national-scale strategies for limiting the magnitude of future climate change. Successfully addressing climate change in the long run, however, requires encouraging simultaneous actions at many scales and capitalizing on the unique benefits to be found at each different scale of action (as discussed by Sovacool and Brown [2009] and Ostrom [2010]). National policy goals will be implemented through the actions of the private sector, state and local governments, and individuals and households. Federal policy needs to provide a framework within which all of these actors can work effectively toward a shared national goal. Similarly, national policy provides a crucial foundation for the role that the United States needs to play internationally in the global effort to limit climate change. This chapter first discusses how U.S. domestic policies can foster strong emissions-reduction efforts by other countries. It then considers how to devise a division of responsibility between federal and state authorities that will promote experimentation within a consistent national policy framework.

INTERNATIONAL STRATEGIES

As discussed in Chapter 2, controlling U.S. greenhouse gas (GHG) emissions will not by itself have a decisive impact on global atmospheric GHG concentrations. However, what the United States does about its own GHG emissions is likely to have a major impact on how other countries respond to climate change. Many countries around the world are already taking strong steps toward lowering their GHG emissions. Lack of action by the United States, however, can serve as an excuse for inaction by others; conversely, if the United States acts vigorously, more countries are likely to follow. U.S. action on climate change, therefore, is as important (if not more important) in regard to its impacts on other countries' behavior as its impact on actual U.S. emissions.

Because climate change is a global issue, many aspects of U.S. climate policy have international implications. Chapter 4 discussed the role of international offsets in domestic climate change policy and how a domestic cap-and-trade system could be linked to international climate agreements. Chapter 6 discussed the ancillary benefits for international relations of effective climate change response measures. This section focuses on strategies for engaging effective international participation in efforts to limit the magnitude of climate change. We discuss U.S. policies in relation to international incentives and the institutions needed to achieve climate change objectives. We then consider specific U.S. policies that could be directed toward low- and middle-income countries where emissions are rising most rapidly, and we illustrate the strategic and contingent nature of the policy problem with three scenarios for future action among these countries. Finally, we examine the roles of both competition and cooperation in encouraging international action to limit climate change.

U.S. Policies and International Incentives

The outcome of efforts to limit climate change will depend on other countries' decisions, but the United States has considerable ability to affect other countries' behavior. As a result, the situation involves *strategic interaction*; that is, nations' policies and ability to cooperate with one another depend not only on their own preferences and resources, but on leaders' beliefs about others' preferences, resources, and strategies (Schelling, 1960; Waltz, 1979). For the United States to engage with other countries most effectively, it needs a strategy for crafting policies that affect the incentives faced by other governments.

The incentive problem is profoundly affected by the fact that climate change policy relates to a common pool resource (or "global commons"): a resource that it is difficult or impossible to exclude others from enjoying but that is degraded by use. In this case, the desired global commons is an atmosphere with a lower concentration of GHGs than would otherwise be the case. Common pool resources are not self-managing; promoting sustained cooperation requires formal institutions involving rules and social norms (Ostrom, 1990).

International institutions (such as the United Nations Framework Convention on Climate Change [UNFCCC] and the Kyoto Protocol) provide principles, rules, and practices around which expectations converge. When credible, they affect beliefs about others' behavior, provide focal points for action, and therefore can affect incentives and outcomes even without having coercive power over states. They can facilitate cooperation by reducing the costs of making and enforcing agreements, providing information, enhancing the credibility of statements by governments and other actors, and providing for the delegation of authority (Hawkins et al., 2006; Ikenberry, 2000; Keohane, 1984). Since these institutions rarely have coercive power, however, they depend on favorable configurations of state preferences and are subject to change when prevailing preferences change.[1]

---

[1] Empirical support for the essential role of international institutions in global environmental policy is provided by data from the International Energy Agency (IEA) Database Project (*http://iea.uoregon.edu/*):

Within international institutions, strategies of reciprocity (making benefits conditional on contributions to the common good) are often the most effective way to generate cooperation, as is generally the case in the international trade system (Barton et al., 2007). It is conceivable that trade-related reciprocity could be employed to provide incentives for other countries to pursue effective climate-change policies. Over-reliance on such a strategy, however, could lead to acrimony and open opportunities for self-interested protectionism by import-competing firms and labor in wealthy countries. Other strategies must be employed to provide incentives for action.

Among these other key strategies for international engagement, it remains crucial to continue to push for strong domestic GHG reduction policies. Without clear, transparent U.S. leadership on this issue, any attempts at fostering greater international engagement are likely to be unsuccessful. A sustainable U.S. domestic strategy must be based on broad public consensus about the importance of limiting climate change, and on policies crafted to meet the interests of key stakeholders in both the public and private sectors. In general, "environmental regulations work most effectively when systems of regulation confer tangible benefits upon the regulated" (Oye and Maxwell, 1995).

The Role of Multilateral Institutions

Climate change policy faces an *institutional dilemma.* High-income countries have the capacity to take strong action to reduce GHG emissions, and they have growing incentives to do so as understanding of the potential damages from climate change grows. But such efforts will be insufficient to limit the extent of global climate change without effective action by the low- and middle-income countries with rapidly growing emissions—countries that numerically dominate multilateral institutions including the United Nations (UN) and the UNFCCC.

The largest of these countries, such as China, India, and Brazil, likewise have incentives to act, as they are major contributors to the problem and may be highly vulnerable to climate change impacts. But having secured exemptions from obligations in the Kyoto Protocol, they are now disinclined to sacrifice those bargaining advantages. The smaller low- and middle-income countries, meanwhile, lack incentives to act vigorously, since climate change is a common-pool resource issue, subject to dilemmas of collective action (that is, small contributors to a common-pool resource problem are strongly tempted to free-ride on the efforts of others).

---

since 1990, over 450 international environmental agreements (treaties, protocols, and amendments) have been signed, which is about as many as in the entire prior century.

Low- and middle-income countries find some protection in the procedures of major-ity-run international organizations in which they are numerically preponderant. In contrast, in bilateral negotiations with the United States and other high-income coun-tries, or with international organizations such as the International Monetary Fund and the World Bank, low- and middle- income countries (except for China and India) are at a profound disadvantage. The implication of this for international engagement is that the formal UN-centered process cannot be abandoned, as this would alienate scores of countries and give strong rhetorical ammunition to voices opposing any action by low- and middle-income countries. Yet, in a huge unwieldy international negotiation with many diverse interests, deadlock can set in. For instance, a decade of active nego-tiation was required for the Law of the Sea Treaty, an issue that was neither as complex nor as important for economic growth as is climate change. Roughly 20 countries are responsible for 75 percent of global emissions (WRI, 2010). The institutional dilemma of climate change policy is how to combine continued adherence to a UN-centered framework with assurance that agreements among the major emitters are not pre-vented or undermined by countries whose emissions are small.

The need for global governance institutions is indicated by the fact that governments repeatedly resort to them. But sovereignty and conflicting state interests ensure that most such institutions will lack simple coherent structures. Many international institu-tions and networks have some bearing on climate change, but none are comprehen-sive or authoritative. The complex range of existing global governance institutions reflects the large variety of interests, bureaucratic organizations, and capacities in and among states. This is an inherent characteristic of world politics, not a problem to be solved. Different problems generate different solutions, and different solutions gener-ate different governance arrangements. Since a single, comprehensive, binding global climate change agreement is unlikely, a multiplicity of mechanisms, linked but not characterized by a single coherent architecture, should be expected to persist. This might include, for example, sectoral-based agreements, discussed in Box 7.1.

Even when agreements are reached, however, they will not be implemented automati-cally. Commitments by governments at international meetings may or may not lead to real action, since such commitments are made in the context of a political process that is often decoupled from implementation. Likewise, even elaborate plans made at the international level do not necessarily mean very much in terms of practical implemen-tation, and they are often vague on metrics for assessment. In general, plans are a lot easier to produce than to implement, as experience with the Kyoto Protocol shows.

In the context of international negotiations, credibility is at a premium. The history of relationships between high- and low-income countries is replete with unfulfilled

---

BOX 7.1
Sectoral Agreements

As discussed in Chapter 4, in the absence of a comprehensive global climate agreement, measures to limit GHG emissions imposed only in high-income economies raise the prospect that a GHG-emitting industry could simply relocate to an emerging-economy country. As a consequence, the international competitiveness of industry in high-income countries could be damaged, and the emissions leakage involved in shifting the location of emissions would reduce the ultimate effectiveness of the policy (see Chapter 4 for further discussion of emissions leakage). Anticipation of such effects would generate political opposition to effective regulation in the high-income countries.

One way to respond to such concerns is to establish global agreements governing relatively well-defined industrial sectors, such as steel and aluminum production, auto manufacturing, and air transport. Such agreements could involve all countries, particularly emerging economies, and would therefore help address issues of competitiveness and leakage. Unlike a comprehensive carbon-pricing system, however, a complex of sectoral agreements would not equalize the marginal costs of emissions control within a jurisdiction.

Sectoral agreements should therefore be seen as a potentially useful interim supplementary practice to limit emissions, rather than as a viable long-term substitute for agreements based on economy-wide caps. Nevertheless, as long as a deep comprehensive agreement involving all countries remains unattainable, sectoral agreements could play a useful role in limiting emissions. The opportunities and challenges of expanding such efforts into more widespread mandatory agreements have been explored in a number of recent analyses (Pew Center, 2007; WBCSD, 2009; WRI, 2007; IEA series at *http://www.iea.org/subjectqueries/sectoralapproaches.asp* [accessed September 17, 2010]).

---

promises. Independent assessment is essential for credibility. As emphasized in *ACC: Informing an Effective Response to Climate Change* (NRC, 2010b), such assessment requires valid standardized methodologies for GHG emissions reporting and verification. Measurement and verification at the international level are particularly difficult due to the absence of institutions with legal powers to inspect sites and sovereignty concerns that limit expansion of such powers. Agreement on valid verification methodologies does not necessarily ensure compliance, if states selling emissions allowances have incentives to cheat on the rules. Since centralized enforcement is difficult or impossible in world politics, constructing a system of buyer liability—modeled on bond markets in which securities of varying quality sell at varying prices—could generate decentralized monitoring and market-generated incentives for compliance (Keohane and Raustiala, 2009). The essential point is that, because enforcement of any international agreement is not likely to be centralized, states must have incentives to implement their own commitments.

Varying perspectives and institutional fragmentation suggest that reaching a meaningful comprehensive agreement on climate change is likely to be a long-term process. Binding global GHG emissions limits cannot be imposed by high-income countries. In the near term, some combination of incomplete international agreements and more specific limited-membership or bilateral agreements is likely. The United States should therefore seek to use the UN process to generate multilateral agreements insofar as feasible but, failing effective global action, should rely also on agreements among major emitters and bilateral accords. In some cases, two countries, or a small set of countries, may have common interests and expertise not shared by others; for example, China and the United States both have interests in developing carbon capture and storage technology. In other cases, if a small group of states forms a "club" that generates benefits exclusive to members, as the European Union has, other states may develop interest in joining rather than holding out for a more favorable deal (Downs et al., 2000).

The United States is already a party to numerous bilateral and multilateral arrangements focused on limiting climate change, including, for instance, the Major Economies Forum on Energy and Climate, the Asia-Pacific Partnership on Clean Development and Climate, the G-8 + 5 Climate Change Dialogue, the Methane to Markets Partnership, the Renewable Energy and Energy Efficiency Partnership, the U.S.–China Partnership on Climate Change, the U.S.–India Nuclear Partnership, and many other activities focused on advancing specific low-carbon energy technologies (e.g., CCS, hydrogen, Gen IV Nuclear, fusion energy). The panel supports continued U.S. involvement and leadership in such efforts, in conjunction with attempts to negotiate appropriate global agreements under the UN.

As discussed in Chapter 3, there is also strong motivation for forging an international agreement to control hydrofluorocarbons (HFCs), greenhouse gases which are the main class of chlorofluorocarbon replacement compounds. Controlling HFCs through the Montreal Protocol may be the best strategy, because the expertise and support structures already existing within that framework could easily be adopted for the applications in which HFCs are used. This also makes sense because HFCs (except by-product HFC-23) are ozone-depleting substance substitutes and, thus, logically fall under the Montreal Protocol's mandate.

In the long run, reaching a meaningful, binding global accord for addressing climate change will depend both on U.S. actions at home and on effective U.S. diplomacy that takes account of the perspectives of other countries. Imagination and resolve will be required, both to identify effective incentives for widespread participation and to ensure rigorous assessment and enforcement of national commitments. All are requisite elements of a credible global solution.

Distinctive Perspectives of Low- and Middle-Income Countries

As discussed earlier, a significant fraction of the emissions reductions required to stabilize global atmospheric GHG concentrations must take place in low- and middle-income countries whose emissions are rapidly increasing. U.S. policy makers will need to understand the positions of leaders and publics in these countries. Some key points to consider include the following:

- Low- and middle-income countries are focused on the need for economic growth, and governments currently in power are unlikely to survive if they fail to produce acceptable rates of growth for their own economies. Hence, these governments have strong incentives not to accept any measures—such as emissions taxes or binding caps on emissions—that could significantly reduce their rates of economic growth. For them to agree to binding targets there would have to be "substantial evidence both that low-carbon growth is feasible and that there will be substantial external technological and financial assistance along the way" (Stern, 2009).
- Low- and middle-income countries often have low and uneven administrative capacity, which makes it difficult to implement complex policies. For a detailed discussion of this issue as applied to India, for example, see Rai and Victor (2009).
- Low- and middle-income countries, by definition, have large numbers of people living below the poverty line (as defined by the World Bank), and their GHG emissions *per capita* are much lower than that of high-income countries. Historically (between 1850 and 2000), the high-income countries emitted more than three times as much CO_2 (from fossil fuels and cement) than the low- and middle-income countries. Equity considerations thus underlie the common demand that high-income countries take the first serious measures to deal with climate change, and that low- and middle-income countries receive compensation in return for their own actions.
- The leaders and publics in some low- and middle-income countries are profoundly suspicious of the motivations and actions of the high-income countries, and they are worried that the burdens of adjusting to climate change will be imposed on them, as reflected in financial and trade policies.

- As discussed in Chapter 6, certain actions that would be beneficial from a climate standpoint—such as limiting black carbon emissions from domestic stoves and diesel engines—would also create substantial health benefits for the countries concerned. Low- and middle-income countries have self-interest incentives to reduce these pollutants.
- Most low- and middle-income countries want to avoid being considered outliers in world politics by failing to join open international institutions. The concept of "common but differentiated responsibilities" embodied in the Kyoto Protocol grew from this interest in belonging to legitimate international institutions but not having to accept onerous obligations. It is likely that most low- and middle-income countries will seek to keep this concept as the defining principle of a global climate strategy.
- From the perspective of the low- and middle-income countries, the United States and other high-income countries will be credible in their demands for action only if they take effective action to limit their own GHG emissions, and if they use international institutions for addressing the problem rather than relying solely on bilateral negotiations or ad hoc groupings.

The different experiences and perspectives between low- and middle-income countries and most high-income countries mean that irreconcilable frameworks for analysis and rhetoric persist. For example, framing climate change as a global commons problem conflicts with framing it in terms of equity. Future-oriented problem-solving perspectives conflict with orientations that seek compensation for past injustices.

Although it is important to appreciate the distinctive common perspectives that exist among low- and middle-income countries, one should not lose sight of the substantial economic, political, and institutional differences among these countries. Some are democracies; some are not. Some have substantial administrative competence to regulate GHGs; others have difficulty implementing any complex regulations. Hence, specific U.S. policies directed toward low- and middle-income countries will have to be carefully differentiated.

The Montreal Protocol and its successor agreements for addressing the problem of stratospheric ozone depletion illustrate that strategies can sometimes be found to overcome the types of problems discussed above; for instance, by setting delayed deadlines for developing countries and establishing a multilateral fund to help them make the transition (Parson, 2003). Yet the cost and complexity of climate change are much greater than for ozone depletion, and forging global cooperation on climate change is a far more difficult problem than the one faced by drafters of the Montreal Protocol.

Contingent Strategies: Three Scenarios

Effective action is always contingent on political realities, and U.S. international strategy must therefore be adaptable to different political configurations evolving over time. The negotiating situation on climate change, as on other international issues, is a "two-level game" (Putnam, 1988): strategies in domestic politics and international politics interact in both directions and must be coordinated. Note, for instance, the useful example provided by Great Britain, which has unconditionally promised a certain level of domestic effort on climate change but also promised further efforts that are conditional on greater action by others (Committee on Climate Change, 2008).

Because it is not possible to predict future developments in global politics, it is instructive to consider a variety of scenarios of the positions that may be taken by key players in a global climate agreement and to evaluate what each scenario might mean for U.S. response strategy. Three scenarios, presented below, all assume willingness by high-income countries to enact vigorous policies to limit the magnitude of climate change, because nothing of consequence can be achieved internationally unless this condition is met. Hence, there is no "business-as-usual" scenario.

Scenario 1: Independent but Coordinated Action

Under the first scenario, all major governments would commit independently to action on climate change, which they would self-finance. Under these conditions, the major task would be coordination: reaching agreement in a situation in which all states had fundamentally compatible interests (Martin, 1992). The issue of carbon pricing could be dealt with through a harmonized set of taxes or (more likely in view of contemporary policies in the European Union and the United States) through either a global cap-and-trade program (Frankel, 2009) or linked national cap-and-trade programs (Jaffe and Stavins, 2009). Large-scale financial transfers would not be required, but joint technology development and transfers would be needed (Newell, 2009). Implementation would be chiefly a national problem, because all countries would be committed to effective action. There would probably need to be an international institution to assess and monitor implementation, providing reassurance to everyone that the burdens were actually being shared. Trade competitiveness issues would not be salient because all countries would undertake comparable obligations.

One could argue that independent action would be beneficial for low- and middle-income countries. For instance, it is noted in Chapter 6 that measures to limit climate change can have substantial agricultural and health co-benefits for these countries.

This scenario, however, seems unlikely to be politically realistic within the next few decades. It would require costly efforts by low- and middle-income countries that many are disinclined even to consider, and such efforts would have to be undertaken by governments that are short of material and administrative resources. In the long run, however, if low- and middle-income countries eventually become more prosperous, independent action could become more realistic.

Scenario 2: A Global Regime with Financial Transfers

To reach a serious international agreement under the UNFCCC, with meaningful action by low- and middle-income countries, some financial transfers from richer to poorer countries seem essential. Indeed, contemporary proposals from some developing countries demand that burdens of funding climate change response (both limiting and adaptation) must be undertaken almost exclusively by high-income countries. There is little evidence, however, to suggest that publics and legislatures in high-income countries are willing to directly provide resources at levels that have been demanded, which could involve hundreds of billions of dollars of financial transfers annually. Such demands contrast sharply with the politics of climate change legislation in the U.S. Congress, which has focused on compensating *domestic* firms and individual voters and on appealing to an interest in long-term *national* energy security and job growth. It seems particularly unlikely that the United States would send very large financial resources to China without compensation of some kind, given the fact that China is a major holder of U.S. government securities and a principal economic rival.

Financial transfers could take place in a cap-and-trade regime involving offsets, or in which the caps applied to low- and middle-income countries are arranged to provide substantial headroom (or "hot air"). International offsets could thereby play a positive role by encouraging other countries to adopt sectoral or economy-wide caps in order to have access to the U.S. market. Under these circumstances, the high-income countries would benefit by having to undertake less domestic effort to reach a particular target, and low- and middle-income countries would benefit financially. (Note that the potential pitfalls of international offsets are discussed in earlier chapters.)

Any major financial transfers to low- and middle-income countries would surely have to be tied to technology transfer and other specific measures that help assure the public that their funding is actually contributing to genuine, additional emissions reductions. Whether this international regime involved some sort of new global institution or was built upon established organizations such as the World Bank, the institutional demands would be substantial. Creating a system that could generate effective incentives for compliance and continued commitment would require considerable institutional ingenuity (Kingsbury et al., 2009).

Scenario 3: A Partial Regime Without Financial Transfers

In a third scenario, the high-income countries would go ahead with vigorous programs to limit emissions but would refuse to make large international financial transfers; low- and middle-income countries would refuse to take vigorous action that they would have to finance themselves. This sort of global regime would be very unsatisfactory in terms of emissions limitation worldwide, and it would raise serious issues of competitiveness insofar as producers in low-income countries were exempted from the costs of emissions control measures. (Similar issues would arise in scenario 2 as a result of financial transfers that relieved producers in low-income countries of the full costs of emissions reductions.) Issues of emissions leakage would be particularly serious.

In this scenario, it would be important to identify "win-win" technological advances (for example, improved agricultural methods that reduce emissions without sacrificing yields, or efficient, low-emissions coal-fired power plants) that help low- and middle-income countries to reduce emissions through measures that promote their economic development. In the absence of financial transfers, creative technological solutions and institutions to facilitate technology transfer would be especially important.

Of great value in this regard are international cooperative efforts such as the International Partnership for Energy Efficiency Cooperation (which helps governments identify and implement policies and programs for promoting energy efficiency) and international cooperative research and development (R&D) activities, such as those listed earlier in this section. A few studies have explored optimal policy strategies for fostering this type of international cooperation (e.g., Golombek and Hoel, 2009; Qiu and Tao, 1998).

Competition and Climate Change

Although explicit cooperation is one way to induce effective action, competition can also generate measures to help limit the magnitude of climate change. Governments are increasingly coming to understand the opportunities for economic growth that can result from playing a leadership role in the transition to a new "green" industrial era. Private-sector and government actions can change rapidly when climate change response is seen as essential for economic competitiveness. China, for example, is actively seeking to maintain growth, create jobs, and increase the competitiveness of its industries by becoming a leader in technologies to conserve energy and reduce GHG emissions. Japan and many EU countries have aggressively moved in this direction as well. Active U.S. engagement in this "race to the top" will both enhance our own future economic growth and motivate other countries to continue moving ahead.

Border Tax Adjustments (BTAs) provide one strategy for inducing more action by reluctant countries and for strengthening the position of negotiators for countries with large potential markets. On the other hand, BTAs are potentially dangerous because (if applied unilaterally) they would be subject to interest-group pressures. And if such measures become protectionist, other countries could retaliate, damaging the international trading system.

A joint report from the World Trade Organization (WTO) and the UN Environment Programme (UNEP and WTO, 2009) suggests that such border adjustments could be legal under WTO rules if two conditions are met: (1) there is a close connection between the means employed and a climate change policy that is either necessary for the protection of human, plant, or animal life or health or relating to the preservation of exhaustible natural resources; and (2) the measure is applied in nondiscriminatory ways that do not serve as "a disguised restriction on international trade." A WTO Appellate Body also insisted that administrative due process be respected: Countries that are the target of environmentally related trade measures must be consulted. Measures to address climate change would meet the first criterion and, if properly applied, could meet the second as well.

The implication is that BTAs could be a valuable part of a climate change policy portfolio, but only if they are firmly established within the WTO-centered international trade regime. Establishing nondiscriminatory BTAs could help to reassure domestic producers about competitiveness, prevent emissions leakage, and at the same time encourage more vigorous action by other countries. They could strengthen the positions of negotiators for countries seeking to take global leadership on climate change. But these policies must conform to WTO law, both substantively and in terms of due process.

As a general strategy then, the United States should strive to promote cooperative measures to limit the magnitude of climate change but should also be alert to ways in which competition can sometimes foster desirable results. A judicious combination of cooperation and competition can generate strong incentives for effective action.

KEY CONCLUSIONS AND RECOMMENDATIONS

Even substantial U.S. emissions reductions will not, by themselves, substantially alter the rate of climate change. Although the United States is responsible for the largest share of historic contributions to global GHG concentrations, all major emitters will need to ultimately reduce emissions substantially. However, the *indirect* effects of U.S. action or inaction are likely to be very large. That is, what this nation does about its own GHG emissions will have a major impact on how other countries respond to the climate change challenge; without domestic climate change limiting policies that are credible to the rest of the world, no U.S. strategy to achieve global cooperation is likely to succeed. Continuing efforts to inform the U.S. public of the dangers of climate change and to devise cost-effective response options will therefore be essential for both global cooperation and effective national action.

The U.S. climate change strategy will need to be multidimensional, operating at multiple levels. Continuing attempts to negotiate a comprehensive climate agreement under the UN Climate Change Convention are essential to establish good faith and to maximize the legitimacy of policy. At the same time, intensive negotiations must continue with the European Union, Japan, and other Organisation for Economic Co-operation and Development countries, and with low- and middle-income countries that are major emitters of, or sinks for, GHGs (especially China, India, Brazil, and the former Soviet Union countries). These multiple tracks need to be pursued in ways that reinforce rather than undermine one another. It may be worthwhile to negotiate sectoral agreements as well, and GHGs other than CO_2 should be subjects for international consideration. In such negotiations, the United States should press for institutional arrangements that provide credible assessment and verification of national policies around the world, and that help low- and middle- income countries attain their broader goals of sustainable development.

Competition among countries to take the lead in low-GHG technology industries will play an important role in stimulating emissions-reduction efforts, but strong cooperative efforts will be needed as well. Sustaining large direct government-to-government financial transfers to low-income countries may pose substantial challenges of political feasibility, but large financial transfers via the private sector could be facilitated via a carbon pricing system that allows international purchases of allowances or offsets. There is a clear need for innovative, cooperative scientific and technological efforts to help low- and middle-income countries limit their emissions. To provide leadership in these efforts, the United States needs to develop and share technologies that not only reduce GHG emissions but also help advance economic development and reduce local environmental stresses.

BALANCING FEDERAL WITH STATE AND LOCAL ACTION

Congress faces complex questions about how to harness the current interest and efforts of state and local authorities while striking an appropriate balance between national policy and state/local regulatory autonomy. This section analyzes how to approach these issues and provides key examples of policy areas that are likely to raise significant questions about overlaps between federal and state policies.

Many states and localities have enacted far-reaching climate policies over the past several years. Researchers estimate that over 50 percent of Americans live in a jurisdiction that has enacted a GHG emissions cap (Lutsey and Sperling, 2008). Figure 7.1 shows the states that have adopted emissions caps and targets, many of which are more aggressive than those being proposed in recent national legislation (i.e., H.R. 2454). In addition, 29 states that contain more than half of the U.S. population have enacted renewable portfolio standards (*http://www.pewclimate.org/*). Fifteen states, representing 30 percent of the country, have indicated their intent to follow California GHG emissions standards for automobiles.[2] The Regional Greenhouse Gas Initiative (RGGI) is already up and running among many northeastern states. California has adopted an ambitious plan to implement its economy-wide emissions cap, passed legislation requiring its localities to incorporate GHG emissions targets into land-use planning, and prohibited importation into the state of electricity that is more GHG-intensive than efficient natural gas facilities. Several states have adopted performance emissions standards for large GHG emitters.

The scope and magnitude of these state programs is potentially enormous, although significant questions remain about whether many of the state programs are more aspirational than real. Nevertheless, if states with emissions caps are successful in achieving their most ambitious targets (and assuming no emissions leakage), the cumulative emissions reductions would approach the 2020 national emissions-reduction targets proposed in H.R. 2454.

Significant growth has also occurred in the scope of climate change response actions occurring at the local/urban level. More than 1,000 mayors have joined the U.S. Conference of Mayor's Climate Protection Agreement, vowing to reduce GHG emissions in their cities below 1990 levels (Figure 7.2). We cannot assess the progress these cities are making in meeting their proposed goals, but there is clearly a great interest and capacity for municipal-scale action to influence many activities and planning deci-

---

[2] The announcement by the Obama Administration that the federal government will adopt fuel economy and GHG emissions standards as tough as California's means the entire country will now be covered by a uniform national standard.

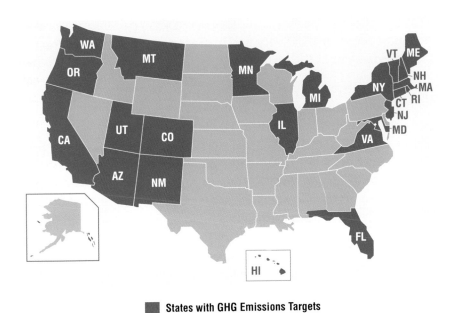

States with GHG Emissions Targets

FIGURE 7.1 States that have adopted emissions caps and targets. SOURCE: Pew Center for Global Climate Change.

FIGURE 7.2 Cities participating in the U.S. Mayor's Climate Protection Agreement (*http://www.usmayors. org/climateprotection/ClimateChange.asp*, accessed September 17, 2010).

sions that affect GHG emissions (e.g., land-use and zoning decisions, infrastructure investment, municipal service delivery, and management of schools, recreation areas, municipal buildings, etc.).

The role of states and localities will also be very important in the implementation and enforcement of many national policies. In most states, for example, cities and counties are responsible for enacting and enforcing building standards. If Congress chooses to enact national building standards, it will still need to rely on localities to ensure that the standards are actually implemented and enforced. If Congress mandates performance standards for certain GHG sources (such as new coal-fired power plants), state-level environmental agencies will be responsible for issuing permits consistent with those standards. Most energy-efficiency building programs are administered at the state and local levels, so if Congress funds large energy-efficiency programs, it will need to rely on subnational jurisdictions to implement those programs.

Encouraging regulatory flexibility across jurisdictional boundaries increases our nation's opportunities to gain valuable experience from state-level experimentation. A key way to ensure this flexibility is for Congress to avoid preempting state and local authority to regulate GHG emissions more stringently than federal law unless there is a strong policy justification for doing so. As a general rule of thumb, we suggest that state action should be preempted only if it is likely to shift significant externalities onto out-of-state residents and consumers—for instance, through leakage of emissions outside the state's borders. Another preemption consideration is the potential burden multiple state and local standards can place on business interests that operate in multiple states with overlapping and/or inconsistent regulations.

Rather than using its preemptive powers to limit state and local regulatory policy, Congress could require states and localities to regulate through the establishment of minimum national standards in areas such as renewable portfolio standards and building standards. This is a commonplace approach in many federal environmental statutes, including the Clean Air Act (CAA) and the Clean Water Act. Minimum national standards may be particularly appropriate for policies traditionally within the purview of states and localities that promise large, cost-effective emissions reductions—such as building efficiency standards.

In considering the appropriate roles that states and localities should play in climate policy, it is also important to consider the way in which various judicial doctrines can preempt state and local action, even where Congress has not passed legislation expressly intended to preempt. When adopting GHG emissions-reduction policies, Congress should make it clear when it intends states and localities to retain independent regulatory authority. Finally, unless there is strong reason, states that have taken early action to reduce GHG emissions should not be penalized by the adoption of federal climate policy.

Below are some examples of policy areas that are likely to raise significant questions about overlapping federal and state policy.

Example: Cap and Trade

Chapter 4 contains a detailed discussion of the policy design challenges associated with a cap-and-trade system. In that discussion, it is noted that one major regional cap-and-trade program (RGGI) is already up and running; other areas of the country are cooperating to possibly establish additional programs (e.g., the Western Climate Initiative). These programs vary in their coverage, with RGGI limited to CO_2 emissions from the power sector, while the Western Climate Initiative has at least initially recommended multisector coverage. Two important questions about the relationship between state and regional programs and federal law arise if the federal government adopts a cap-and-trade program.

First is the question of preemption. State cap-and-trade programs may cover more sources or impose a tighter cap than a comparable federal program. If federal legislation contains provisions that significantly undermine its effectiveness (e.g., overly generous safety valves or borrowing provisions), then a more stringent state program could compensate through additional emissions reductions within the state. With a federal cap in place, however, more stringent state caps may not actually produce any net decrease in U.S. emissions. Rather, this may simply help the rest of the country meet the federal cap more easily, because fewer reductions will be required outside of those states with stringent caps. Thus, state cap-and-trade programs may simply redistribute the allocation of emissions reductions and limit the freedom of a carbon market to distribute reductions efficiently (McGuinness and Ellerman, 2008).

Congress could allow states to effectively lower the federal emissions cap by permitting them to influence the overall number of allowances in the market. For example, a state with a cap-and-trade program could be allowed to prohibit its sources from selling unused allowances on the open market. Similarly, a state could auction allowances and then use revenue from the auction to purchase and retire federal allowances.

If state cap-and-trade programs are not preempted (i.e., are allowed to continue), and in-state sources are restricted more stringently than out-of-state sources, this may increase the overall costs of emissions reductions within the state. And if fewer sources are subject to a cap-and-trade program, this may increase volatility in the market for allowances (McGuinness and Ellerman, 2008). It may also increase compliance costs for companies that operate nationwide, if they are required to meet multiple state program requirements. Conversely, if state or regional cap-and-trade programs are preempted, states may have difficulty meeting their own stated emissions targets (if they are more stringent than the targets proposed in recent federal legislation).

Another incentive for states to enact cap-and-trade programs more stringent than a federal program is to maintain funding streams for supporting state-level programs, for instance, related to energy efficiency, renewable energy, and R&D. By shifting costs onto in-state sources, more stringent state programs can lower allowance prices for the rest of the country, conceivably making adoption of a national cap-and-trade program more politically feasible.

Whether Congress should preempt state cap-and-trade programs is, therefore, a matter of debate. On the one hand, there is good reason for Congress to avoid preempting if states with more stringent policies will themselves incur the costs of those policies rather than exporting them. On the other hand, a principal objective of a cap-and-trade scheme is to allow market forces to encourage the cheapest emissions reductions. If state programs coexist with a federal program, they risk making emissions reductions more expensive and increasing compliance costs for companies operating in multiple states.

An additional consideration is how to avoid punishing states that have taken early action to reduce GHG emissions. If a federal program preempts state or regional cap-and-trade programs already in existence and does not provide credit for state program allowances, the value of those allowances will decline to zero (a decline that could occur as soon as federal legislation passes but well before the federal program takes effect). Emissions sources with banked allowances would begin to use those allowances, driving allowance prices down. To avoid this scenario, Congress could either permit state allowances to be transferred into the federal program, and reduce the number of federal allowances by an equal amount, or it could permit state allowances in addition to the federal allowances that are allocated. The latter will, however, expand the federal cap by the total amount of the state allowances, which could compromise efforts to reach national emissions-reduction goals (McGuinness and Ellerman, 2008).

Another significant issue arises for states operating under the RGGI program. Several RGGI states auction off their allowances and use the revenues for energy efficiency and renewable resource promotion. If RGGI is preempted and no accommodation is made for the revenue loss, these programs will suffer a serious setback. Congress may thus wish to ensure that the revenues from RGGI energy efficiency and renewable resource programs are replaced by federal revenue.

Example: Clean Air Act and Preemption

The U.S. Supreme Court, in *Massachusetts v. EPA*, made clear that GHG emissions are "pollutants" under the Clean Air Act and that the EPA must determine whether these pollutants endanger public health and welfare. The EPA has now made such a determination (in its so-called endangerment finding). If Congress adopts GHG emissions legislation, it must clarify the extent to which the CAA will continue to cover the regulation of GHG emissions. Even if Congress amends the CAA to remove GHG emissions regulation from the statute, open questions related to the states remain.

First, for instance, is the question of whether California's special authority to regulate mobile source emissions under the CAA should continue to cover GHG emissions. For 40 years, the state has played a pioneering role in mobile source regulation, including adopting the GHG emissions standards that form the basis of the Obama Administration's establishment of a national Corporate Average Fuel Economy standard. The history of the exercise of California's authority has generally followed a consistent pattern: California first adopts mobile source standards that are more stringent than federal standards, and then, if the standards are successful, the federal government follows suit. Researchers have concluded that this pattern has significant national benefits. California can engage in regulatory experimentation; other states can follow California's lead but are not required to do so. Frequently, the California experience has demonstrated that more stringent regulations can accomplish significant pollution reductions at much lower cost than the auto industry predicted. For instance, the current "Tier II" federal auto standards are a direct result of California's experience in adopting stringent fleet standards at far lower cost than initially predicted (Carlson, 2008), resulting in impressive pollution reductions (e.g., NO_X emissions have dropped more than 99 percent below 1970 levels).

The special status granted to California also has the benefit that it limits the number of regulatory standards manufacturers must meet to two, as opposed to the potential for 50 separate state standards, thus balancing the benefits of regulatory experimentation with the advantages of a uniform regulatory standard. For all these reasons, it seems justified to argue that, as long as California's special regulatory authority does not interfere with advancing federal efforts, it should be allowed to continue. For instance, California's authority to enact more stringent GHG emissions standards after 2016 (when the newly announced national standards are in full effect) should be made clear in any national climate change legislation.

Second is the question of whether states should be allowed to adopt performance standards for stationary sources of GHG emissions. Currently California, Montana,

Oregon, and Washington have standards that in various forms essentially prohibit the construction and/or purchase of electricity generated from coal absent some form of sequestration. (Similar standards established in Massachusetts and New Hampshire were superseded by RGGI.) The CAA makes clear that, in regulating pollutants from stationary sources, states have the authority to exceed federal standards. So if the CAA is amended to limit the regulation of GHG emissions, Congress should be clear about whether states will retain independent authority to regulate.

A few reasons have been proposed for why states should be allowed to impose their own stationary source standards. First, states regulating in excess of federal standards would likely shift abatement costs from out of state to in-state sources In other words, the regulating state will bear the costs of its regulatory decision (McGuinness and Ellerman, 2008). Second, performance standards can have technology-forcing effects, as occurred with California's mobile source standards for conventional pollutants. Although performance standards may reduce economic efficiency (by forcing GHG reductions in particular places rather than allowing market forces to determine the cheapest reductions), this is offset by the innovations that can result from regulatory experimentation and flexibility.

The strategy of not preempting state programs has precedents. In enacting the national cap-and-trade program for SO_2 emissions, Congress did not preempt more stringent state regulation; to this day, several states continue to regulate SO_2 emissions more stringently than federal standards. Similarly, when the Bush Administration adopted a cap-and-trade program for mercury (known as the Clean Air Mercury Rule, subsequently struck down by a federal appellate court), states retained the authority to regulate mercury more stringently than federal law; again, several did so.

Example: Building Standards

Building standards have traditionally remained within the purview of local governments and states. Adoption of new and more stringent standards (such as Leadership in Energy and Environmental Design) has been uneven across jurisdictions. Given the substantial benefits of energy efficiency savings, Congress may wish to engage in "floor preemption," in which states are required to adopt minimum energy-efficiency standards for buildings but are also allowed to exceed those minimums. This is a commonplace approach in federal environmental statutes, including the CAA and the Clean Water Act.

If Congress does enact minimal national building codes, however, it should ensure that states and/or localities responsible for implementing standards have sufficient resources and personnel to rigorously enforce standards. These resources could potentially be funded by allowance revenue (if allowances are auctioned). Absent effective enforcement, national building standards could be significantly undermined just as recent evidence has suggested uneven-to-nonexistent enforcement by some states under the Clean Water Act. Enforcement issues are addressed in more detail in Chapter 8.

KEY CONCLUSIONS AND RECOMMENDATIONS

Given the important role that state and local actions are currently playing in U.S. efforts to reduce GHG emissions, the many ways in which states and localities will be needed for implementing new federal initiatives, and the value of learning from policy experimentation at subnational levels, Congress should carefully balance federal and state/local authority and promote regulatory flexibility across jurisdictional boundaries, with consideration of factors such as the following:

- The need to avoid preempting state/local authority to regulate GHG emissions more stringently than federal law ("ceiling preemption") without a strong policy justification;
- The need for any new GHG emissions-reduction policies to clearly indicate whether a state retains regulatory authority (given various judicial doctrines that can preempt state and local action even without express preemption language from Congress);
- The need to ensure that states and localities have sufficient resources to implement and enforce any significant new regulatory burdens placed on them by Congress (e.g., national building standards); and
- The importance of not penalizing or disadvantaging states (or entities within the states) that have taken early action to reduce GHG emissions.

CHAPTER EIGHT

Policy Durability and Adaptability

The complexity of efforts to reduce greenhouse gas (GHG) emissions and the breadth of change necessary to achieve large reductions mean that the portfolio of U.S. climate change policies will need to be continually evaluated and revised as we gain new information and experience. Devising an optimal long-term policy portfolio from the outset is unlikely, because many of the policies that need to be enacted have never been tested at the national level, and also because climate policies cut across virtually every sector of the economy and are likely to interact in unexpected ways. It is therefore crucial that any major climate change policies enacted by Congress include flexible, adaptable mechanisms for responding to new information.

At the same time, however, it is crucial to ensure that the policies enacted are durable—that is, properly enforced and resistant to subsequent distortion and undercutting. There are inherent tensions between these goals of adaptability and durability, and it will be an ongoing challenge to find an appropriate balance between them. To help meet this challenge, it is imperative that processes be established at the outset for generating and disseminating to policy makers a broad array of information about relevant scientific and technological developments and about the effectiveness and costs of existing policies. These concepts are discussed further in the following sections.

POLICY STABILITY, DURABILITY, AND ENFORCEMENT

Climate policy must be sufficiently durable to last for the decades that will be required to achieve a long-term transition to a low-carbon economy. Both the types of policy instruments chosen and the ways in which these policies are implemented will affect their durability. There has been a great deal of variation of the durability of policy reforms in U.S. experience. Understanding the sources of this variation is extremely important in order to increase the probability that legislation to limit the magnitude of climate change will be sustainable in the long run.

Patashnik (2008) studied the question of reform sustainability in the context of the Tax Reform Act of 1986 and the "Freedom to Farm" Act of 1996. Both acts kept in place existing institutional structures and well-organized interest groups ready to whittle away or even reverse the policy reforms. The Tax Reform Act, hailed as a landmark at the time, was virtually nullified over the next 20 years by legislation enacting exceptions and changes. The market-oriented reforms of the "Freedom to Farm" Act were largely reversed 6 years after enactment.

In contrast, airline deregulation enhanced efficiency, destroyed the old institutional structure (centered on the Civil Aeronautics Board), and fostered a market-led reorganization of the industry. Old carriers entered bankruptcy; new low-cost carriers, empowered by a deregulated environment, were created. It soon became clear that it would be pointless to try to reverse deregulation, discouraging efforts to do so.

Most directly relevant to climate change policy, Title IV of the 1990 Clean Air Act (CAA) Amendments, establishing a cap-and-trade system for SO_2, has transformed this policy area and become self-sustaining, with no prospect of returning to command-and-control regulation. It has persisted for at least three reasons. First, it achieved its environmental objective and did so in a cost-effective fashion. Second, firms preferred to make their own decisions and face emissions prices rather than have command-and-control regulations imposed on them. Third, Title IV created what were in effect (if not strictly in law) property rights in emissions allowances, which gave their holders incentives to support continuation of the program. The banking provisions of Title IV were politically important, because firms that had banked allowances had particularly strong incentives to favor continuation of the system (Patashnik, 2008).

Successful reforms create or rely on government structures that are designed to support the reforms. They change the agents (or coalitions of agents) that dominate policy implementation. Specifically, reforms are sustainable when the major players have interests in their continuation. A key lesson for ensuring policy durability is to create a constituency that benefits from the policies and therefore has a vested interest in maintaining them. As explained in Chapter 4, this rationale may, in some cases, provide a basis for preferring a cap-and-trade scheme, which creates property rights in holders of emissions allowances. Similarly, regulatory policies that spur technological innovation can create a constituency for ongoing federal regulation, something that has occurred previously, for instance, with hazardous waste regulation and reformulated gasoline requirements (Lazarus, 2004; Revesz, 2001). For example, if Congress provides ambitious incentives and funding to stimulate the development of carbon capture and storage, firms that have developed the needed technological advances are likely to advocate federal policies that require the use of such technology.

Although policy instrument choice can enhance policy stability, even the best-crafted legislation can face significant impediments that undermine its support in the implementation stage. For instance, although many of the major federal environmental statutes enacted in the 1970s have achieved significant environmental success, the record of implementation and enforcement of the statutes is frequently marred by delays and failures to enforce. Even today, for example, numerous areas of the country remain in violation of key provisions of the CAA. Similarly, widespread violations of the Clean Water Act have been reported to occur regularly across the country, and yet the Environmental Protection Agency (EPA) and state authorities have failed to take enforcement action against even egregious violators (Duhigg, 2009).

Given the need to move quickly to cut GHG emissions, special attention should be paid to ensuring that emissions-reduction policies are well implemented and properly enforced. Many programs have been significantly hampered in their implementation because the implementing agency lacks the appropriate level of staff necessary to carry out its responsibilities. This was likely the case in the early 1970s, when the EPA was tasked with simultaneously implementing a number of new and highly complex statutes. Similarly, the Department of Energy (DOE) appears to have been understaffed for years in the appliance efficiency standards program, leading to long delays in the issuance of updated standards (GAO, 2007). At a minimum, Congress should ensure that agencies have appropriate levels of staffing necessary to implement complex policies.

Staffing concerns also relate to the question of policy durability discussed above. In order to reduce the likelihood that policies are undermined by special interests or waning public attention, Congress could help buffer the policies from the annual appropriations process and ensure adequate funding levels by providing agencies with self-financing for staffing purposes (Lazarus, 2004). The agency administering cap-and-trade offsets, for instance, could be authorized to charge a fee to offset users in order to fund the staff needed for effectively overseeing offset administration.

It is worth noting that Congress has sometimes taken an even stronger approach when concerned that interest group opposition to regulatory activity might delay or obstruct action. For example, when Congress amended the Resource Conservation and Recovery Act in 1984, it set statutory deadlines for the EPA to issue pretreatment standards for various categories of hazardous waste before the waste could be disposed of on land. The EPA had previously violated deadlines in issuing regulations, in part because of industry lawsuits. Congress was therefore wary of further delay and included this regulatory "hammer" provision to encourage industry to cooperate with the EPA in its development of standards in order to meet the statutory deadline. These sorts of hammer provisions are not always successful,[1] but this example illustrates that Congress has numerous tools at its disposal in implementing complex statutes.

Different policies will have different requirements for ensuring effective implementation and enforcement. Below are two examples associated with complementary policies, which illustrate how effective implementation is needed to ensure that GHG emissions targets are actually achieved.

Example 1: Effective Implementation of Renewable Portfolio Standards

A Renewable Portfolio Standard (RPS) requires electricity retailers to purchase a certain percentage of electricity supplies from renewable sources such as solar and wind. A central issue in ensuring the integrity of an RPS program is preventing the double-counting of renewable resources by more than one electricity retailer, which would significantly undermine the effectiveness of a program. One strategy to avoid double-counting is to prohibit retailers from using voluntary consumer purchases of renewable energy to count toward the RPS.

There will also be a need to ensure that the federal program works in concert with existing state programs. For example, if federal credits are tradable with state credits, this would raise complicated questions if the federal and state programs differed regarding what constitutes an eligible resource. It seems wise, in this case, for the agency designated to administer an RPS to be given regulatory authority to promote flexibility in the administration of state programs while ensuring that efforts to harmonize state and federal programs do not lead to undermining the effectiveness of the RPS.

Another important implementation question is how to track renewable energy credits (which provide a record of renewable energy produced and allow for credit trading) in order to verify them and trace their ownership (Cory and Swezey, 2007). The administering federal agency must be given sufficient resources to design and develop an effective tracking mechanism. It will be useful to draw upon the experience gained by states that have already developed such programs.

For the administering agency to enforce the RPS standard in a meaningful fashion, it also will need authority to investigate and penalize entities that violate the rules of the RPS program, including electricity retailers and renewable generators. Moreover, electricity retailers that fail to meet their RPS obligations should face fines or alterna-

---

[1] In one example, a drop-dead provision in the Clean Air Act aimed at preventing a failure to meet auto emissions standards subsequently had to be amended when no domestic auto manufacturers were prepared to meet the deadline.

tive compliance payments. Finally, the administering agency must be given adequate budgetary resources in order to staff the program effectively.

Example 2: Implementation of Appliance Efficiency Standards

Appliance efficiency standards will likely be part of the national policy portfolio for reducing GHG emissions. The DOE has long had authority to issue appliance standards and is required by statute to periodically reevaluate and reissue new standards for a large number of residential and commercial products. States are generally preempted from issuing their own standards if federal standards are in place, unless the DOE grants a state waiver based on stringent statutory requirements. To date, DOE has missed every congressional deadline set for establishing energy-efficiency standards (GAO, 2007). The Government Accountability Office (GAO) concluded that a principal reason for this failure to comply with statutory deadlines has been the lack of sufficient funding to adequately staff the program. This implies that, at a minimum, Congress needs to provide adequate funding for DOE to meet its statutory obligations in a timely fashion.

Congress may want to take additional measures to strengthen the likelihood that appliance standards will be issued as early as possible, in order to maximize energy savings and consequent GHG emissions reductions. In particular, Congress could make it easier for states to set their own appliance standards. The current waiver language requires a state to demonstrate that more stringent state regulation is necessary to meet "unusual and compelling State or local energy or water interests" that "are substantially different in nature or magnitude than those prevailing in the United States generally" (42 U.S.C. 6297(d)(1)(A) (2007)). To date only one waiver request has been filed (by California), and it was denied. Congress could alter this preemption language by instead allowing for waivers if the proposed new appliance standards are more stringent than federal law. Other possible options include allowing a "California exemption" (as with autos) or allowing for alternative state standards if DOE has missed its statutory deadline for setting new standards.

GENERATING TIMELY INFORMATION FOR ADAPTIVE MANAGEMENT

Congress and the executive branch must remain informed about a wide array of scientific, technical, and economic information related to climate change and to our nation's response strategies. In some contexts, policy makers face a paucity of relevant information (for instance, subnational-level policy makers may lack needed information about location-specific climate change impacts). At the national level, the problem is not necessarily a lack of information, but rather the fact that available information can be scattered and unwieldy to manage and can get lost among the huge array of issues competing for the attention of policy makers.

There are a variety of strategies that could be used to help address such concerns, as discussed in *ACC: Informing Effective Responses to Climate Change* (NRC, 2010b). We suggest, as one example, a process in which the President periodically (e.g., every 2 years) reports to Congress on key developments affecting our nation's response to climate change. This "Climate Report of the President" can be seen as analogous to the Economic Report of the President (prepared annually by the Chair of the Council of Economic Advisers), which sets forth the President's national and international economic policies and reports on the state of the nation's economy.

Another example is the reporting requirement imposed on the EPA in Title VI of the CAA, which requires the agency to report to Congress every 3 years on the concentration and impacts of gases that deplete the stratospheric ozone layer. Placing these sorts of reporting mechanisms at the presidential level (rather than the federal agency level) may pose some risk of politicizing the issues involved, but in the case of climate change policy, this may be essential for ensuring sufficient attention to the issue. In addition, the wide array of national and subnational entities involved in addressing this issue may render an agency-level reporting mechanism impractical.

Regardless of the exact mechanism used, this sort of reporting requirement would serve the purpose of creating a focal point for decisions on climate change and an opportunity for advocates on all sides to attract attention to their criticisms and ideas for policy change. Politicians can use the opportunity to float alternative proposals or call attention to weaknesses in the implementation of current policies. Nothing can force recalcitrant bureaucracies to act, but public scrutiny can motivate them to make decisions or be more energetic about enforcing their own rules. In addition, regulatory mechanisms can be established that force agencies to act upon significant new information that becomes available through the report.

This sort of assessment and reporting process is of course a significant undertaking, which would require staffing and resources from (and coordination among) numerous government agencies. But the process could build upon several existing government mechanisms for periodic reporting on key climate change information—including, for instance, the annual GHG emissions inventory carried out by the EPA, and the U.S. Climate Action report organized by the State Department as input to the United Nations Framework Convention on Climate Change and the annual "Our Changing Planet" report compiled by the U.S. Global Change Research Program. In comparison to these existing efforts, this process would examine a wider base of information pertinent to the effectiveness of U.S. efforts to limit the magnitude of future climate change. For instance, it may include updates on

- national and global emissions trends and their relationship to developments in our understanding of climate change science (including reporting on whether the United States is making sufficient progress toward meeting its GHG budget);
- energy market developments and trajectories;
- the implementation status, costs, and effectiveness of GHG emissions-reduction policies;
- the status of the development and deployment of key technologies for reducing GHG emissions;
- the distributional consequences of emissions-reduction policies across income groups and regions of the country; and
- developments in understanding of climate change impacts and vulnerability to those impacts; updates of adaptation plans and actions under way at federal, state, and local levels.

As discussed above, climate change response policies must have stability, to ensure that investment incentives are maintained, but they will also need to evolve as new information becomes available and as we gain more experience with various policy choices. Lessons can be drawn from other policy areas about how to strike this balance between stable but adaptive regulatory mechanisms.

For example, CAA Section 108 requires the EPA to establish National Ambient Air Quality Standards (NAAQS) for pollutants that "may reasonably be anticipated to endanger public health or welfare." By including within the statute a broad definition of "pollutants" (and setting a scientific threshold for adding new pollutants that is neither negligible nor unreasonably high), Congress has allowed the EPA to add standards for new pollutants as more scientific information becomes available. The EPA used this authority to add a new NAAQS for fine particulate matter (PM2.5) in 1997, almost three decades after the process was established. Likewise, Section 108 specifies a process for revising existing ambient standards as the underlying scientific evidence changes. Again using PM2.5 as an example, the EPA revised the 24-hour NAAQS in 2006 and has periodically revised others, including standards for ozone.

Congress may wish to consider applying similar evolutionary mechanisms to climate legislation. The definition of GHG pollutants, for example, should be broad enough to include any gases subsequently discovered to contribute substantially to climate change. And in any statutes setting performance standards, language should be included that requires the implementing agency to update the standards based on new scientific and technical information. These sorts of regulatory mechanisms can themselves serve as focal points for agency action and for advocates on all sides of the climate debate to encourage agency responsiveness.

Adaptive policy mechanisms can also be applied to market-based policies. For instance, harvesting caps in fisheries around the world are often set annually, rather than having a fixed cap. The annual cap is based on the best available science on fish stocks for that year. Rights to harvest a specific amount of fish each year (akin to emissions allowances in a cap-and-trade program) are then allocated to rights holders as a percentage of the cap (NRC, 1999). Similarly, Congress could consider providing authority to the administrator of a cap-and-trade program to alter the cap on a periodic basis in response to new information about progress in meeting long-term emissions goals, the cost of pricing mechanisms, and changes in available technologies.

An example of another policy evolution mechanism is Japan's Top Runner Program, which uses progress in the commercial development of efficiency technology for vehicles and appliances to set efficiency standards. The program works by using the energy performance of the best available technology on the market to set standards. The standards typically take effect 4 to 8 years after the technology is available commercially. For passenger automobiles, for example, Japan set Top Runner standards in 1999 to improve fuel economy by 22.8 percent by 2010; the targets were met by 2005. Preliminary analysis suggests that the program has been quite effective in moving Japan toward its targets for cutting GHG emissions (METI, 2008; Nordqvist, 2006).

Similarly, statutory deadlines that require the adoption of updated efficiency standards for appliances and automobiles can produce policy evolution. As noted above, however, federal agencies have frequently missed statutory deadlines, so care needs to be taken that the responsible agency has sufficient resources and staffing in place to meet deadlines.

KEY CONCLUSIONS AND RECOMMENDATIONS

The strategies and policies discussed in this report are complex efforts with extensive implications for other domestic issues and for international relations. It is therefore crucial that policies be properly implemented and enforced, and that they be designed in ways that are durable and resistant to distortion or undercutting by subsequent pressures. At the same time, policies will need to be sufficiently flexible to allow for adaptation as we gain experience and understanding. There are inherent tensions between these goals of durability and adaptability, and it will be an ongoing challenge to find a balance between them.

Such efforts require transparent, predictable mechanisms for policy adaptation and processes for ensuring that policy makers receive timely information about scientific, economic, technological, and other relevant developments. A process for periodic collection and analysis of key information related to our nation's climate change response efforts (for instance, in the form of a "Climate Report of the President") would provide a focal point for analysis, discussion, and public attention. The process would be particularly useful if it includes requirements for the responsible implementing agencies to act upon pertinent new information gained through this reporting mechanism.

References

Abrahamse, W., L. Steg, C. Vlek, and T. Rothengatter. 2005. A review of intervention studies aimed at household energy conservation. *Journal of Environmental Psychology* 25(3):273-291.

Alderfer, R. B., and T. J. Starrs. 2000. *Making Connections: Case Studies of Interconnection Barriers and Their Impact on Distributed Power Projects.* National Renewable Energy Laboratory Report NREL/SR-200-28053. Golden, CO: NREL.

Alic, J. A., D. S. Mowery, and E. S. Rubin. 2003. *U.S. Technology and Innovation Policies: Lessons for Climate Change.* Arlington, VA: Pew Center on Global Climate Change.

Altshuler, A. A., and D. Luberoff. 2003. *Mega-projects: The Changing Politics of Urban Public Investment.* Washington, DC: Brookings Institution Press.

American Physical Society. 2008. *Energy Future: Think Efficiency.* College Park, MD: American Physical Society.

Antle, J. M., S. M. Capalbo, S. Mooney, E. T. Elliott, and K. H. Paustian. 2001. Economic analysis of agricultural soil carbon sequestration: An integrated assessment approach. *Journal of Agricultural and Resource Economics* 26(2):344-367.

Appelbaum, E., and P. Alpin. 1990. Differential characteristics of employment growth in service industries. In *Labor Market Adjustments to Structural Change and Technological Progress*, E. Appelbaum and R. Schettkat, eds. New York: Praeger.

Arvai, J., G. Bridge, N. Dolsak, R. Franzese, T. Koontz, A. Luginbuhl, P. Robbins, K. Richards, K. S. Korfmacher, B. Sohngen, J. Tansey, and A. Thompson. 2006. Adaptive management of the global climate problem: Bridging the gap between climate research and climate policy. *Climatic Change* 78(1):217-225.

AWEA (American Wind Energy Association). 2009. *Annual Wind Industry Report.* Washington, DC: AWEA.

Baker, J. S., B. A. Mccarl, B. C. Murray, S. K. Rose, R. J. Alig, D. M. Adams, G. Latta, R. H. Beach, and A. Daigneault. 2009. *Effects of Low-Carbon Policies on Net Farm Income.* Working Paper NI WP 09-04. Durham, NC: Nicholas Institute for Environmental Policy Solutions.

Barker, T., and I. Bashmakov. 2007. Mitigation from a cross-sectoral perspective. In *Climate Change 2007: Mitigation, Contribution of Working Group III to the Fourth Assessment Report of the Intergovernmental Panel on Climate Change*, B. Metz, O. R. Davidson, P. R. Bosch, R. Dave, and L. A. Meyer, eds. Cambridge: Cambridge University Press, 851 pp.

Bartels, L. M. 2008. *Unequal Democracy: The Political Economy of the New Gilded Age.* Princeton, NJ: Princeton University Press.

Bartik, T. J. 1991. *Who Benefits from State and Local Economic Development Policies?* Kalamazoo, MI: W.E. Upjohn Institute for Employment Research.

Barton, J. H., J. L. Goldstein, T. E. Josling, and R. H. Steinberg. 2007. *The Evolution of the Trade Regime.* Princeton, NJ: Princeton University Press.

Basu, R., and B. D. Ostro. 2008. A multicounty analysis identifying the populations vulnerable to mortality associated with high ambient temperature in California. *American Journal of Epidemiology* 168(6):632-637.

Bazerman, M. H., and A. J. Hoffman. 1999. Sources of environmentally destructive behavior: Individual, organizational, and institutional perspective. *Research in Organizational Behavior* 21:39-79.

Beierle, T. C. 1998. *Public Participation in Environmental Decisions: An Evaluation Framework Using Social Goals.* Discussion Paper 99-06, 31 pp.

Bielicki, J. M., and J. C. Stephens. 2008. Public Perception of Carbon Capture and Storage Technology. Cambridge, MA: Belfer Center for Science and International Affairs at Harvard University's John F. Kennedy School of Government. Available at *http://belfercenter.ksg.harvard.edu/files/CCS_Public_Perception_Workshop_Report.pdf.*

Bin, S., and H. Dowlatabadi. 2005. Consumer lifestyle approach to US energy use and the related CO_2 emissions. *Energy Policy* 33(2):197-208.

Blanford, G. J., R. G. Richels, and T. F. Rutherford. 2009. Feasible climate targets: The roles of economic growth, coalition development and expectations. *Energy Economics* 31(2):S82-S93.

Boucher, D. 2008. *Out of the Woods: A Realistic Role for Tropical Forests in Curbing Global Warming.* Union of Concerned Scientists: Washington, DC.

Bosetti, V., C. Carraro, A. Sgobbi, and M. Tavoni. 2008. *Delayed Action and Uncertain Targets: How Much Will Climate Policy Cost? Euro-Mediterranean Centre for Climate Change* (CMCC) Research Paper 46. Italy.

Boyce, J. K., and M. Riddle. 2007. *Cap and Dividend: How to Curb Global Warming While Protecting the Incomes of American Families.* Amherst: Political Economy Research Institute, University of Massachusetts, Amherst.

Boyd, P. W., T. Jickells, C. S. Law, S. Blain, E. A. Boyle, K. O. Buesseler, K. H. Coale, J. J. Cullen, H. J. W. De Baar, M. Follows, M. Harvey, C. Lancelot, M. Levasseur, N. P. J. Owens, R. Pollard, R. B. Rivkin, J. Sarmiento, V. Schoemann, V. Smetacek, S. Takeda, A. Tsuda, S. Turner, and A. J. Watson. 2007. Mesoscale iron enrichment experiments 1993-2005: Synthesis and future directions. *Science* 315(5812):612-617.

Bressand, F., D. Farrell, P. Haas, F. Morin, S. Nyquist, J. Remes, S. Roemer, M. Rogers, J. Rosenfeld, and J. Woetzel. 2007. *Curbing Global Energy Demand Growth: The Energy Productivity Opportunity.* Washington, DC: McKinsey Global Institute.

Brown, M. A., and S. Chandler. 2008. Governing confusion: How statutes, fiscal policy, and regulations impede clean energy technologies. *Stanford Law and Policy Review* 19(3):472-509.

Brown, M. A., and F. Southworth. 2008. Mitigating climate change through green buildings and smart growth. *Environment and Planning A* 40(3):653-675.

Brown, M. A., T. K. Stovall, and P. Hughes. 2007. Tackling climate change in the United States: The potential for greenhouse gas reductions in the buildings sector. In *Tackling Climate Change in the U.S.: Potential Carbon Emissions Reductions from Energy Efficiency and Renewable Energy by 2030,* C. F. Kutscher, ed. Washington, DC: American Solar Energy Society.

Brown, M. A., J. Chandler, M. Lapsa, and M. Ally. 2009a. *Making Homes Part of the Climate Solution.* Oak Ridge National Laboratory Report ORNL/TM-2009/104, Oak Ridge, TN.

Brown, M. A., F. Southworth, and A. Sarzynski. 2009b. The geography of metropolitan carbon footprints. *Policy and Society* 27(4):285-304.

Brown, R., S. Borgeson, J. Koomey, and P. Biemayer. 2008. *U.S. Building-Sector Energy Efficiency Potential.* Berkeley, CA: Lawrence Berkeley National Laboratory.

Brulle, R. J., and D. N. Pellow. 2006. Environmental justice: Human health and environmental inequalities. *Annual Review of Public Health.* Vol. 27.

Buesseler, K. O., S. C. Doney, D. M. Karl, P. W. Boyd, K. Caldeira, F. Chai, K. H. Coale, H. J. W. De Baar, P. G. Falkowski, K. S. Johnson, R. S. Lampitt, A. F. Michaels, S. W. A. Naqvi, V. Smetacek, S. Takeda, and A. J. Watson. 2008. Environment: Ocean iron fertilization: Moving forward in a sea of uncertainty. *Science* 319(5860):162.

Bullard, R. D. 2000. *Dumping in Dixie: Race, Class, and Environmental Quality,* 3rd ed. Boulder, CO: Westview Press.

Burtraw, D., A. Krupnick, K. Palmer, A. Pul, M. Toman, and C. Bloyd. 1999. *Ancillary Benefits of Reduced Air Pollution in the U.S. from Moderate Greenhouse Gas Mitigation Policies in the Electricity Sector.* Washington, DC: Resources for the Future.

Burtraw, D., R. Sweeney, and M. Walls. 2008. *The Incidences of U.S. Climate Policy: Where You Stand Depends on Where You Sit.* Discussion Paper RFF-DP-08-28. Washington, DC: Resources for the Future.

Burtraw, D., K. Palmer, and D. Kahn. 2009. *A Symmetric Safety Valve.* Washington, DC: Resources for the Future.

Busch, J., B. Strassburg, A. Cattaneo, R. Lubowski, A. Bruner, R. Rice, A. Creed, R. Ashton, and F. Boltz. 2009. Comparing climate and cost impacts of reference levels for reducing emissions from deforestation. *Environmental Research Letters* 4(4):044006.

California Energy Commission. 2005. *Options for Energy Efficiency in Existing Buildings.* Commission Report. CEC-400-2005-039-CMF. Available at *http://www.energy.ca.gov/2005publications/CEC-400-2005-039/CEC-400-2005-039-CMF.PDF.* Accessed on July 27, 2010.

Capek, S. M. 1993. The environmental justice frame: A conceptual discussion and an application. *Social Problems* 40(1):5-24.

CARB (California Air Resources Board). 2008. *Climate Change Proposed Scoping Plan: A Framework for Change, October 2008, Pursuant to AB 32, the California Global Warming Solutions Act of 2006.* Available at *http://www.arb.ca.gov/cc/scopingplan/document/scopingplandocument.htm.*

Carlson, A. E. 2008. *Iterative Federalism and Climate Change.* UCLA School of Law Research Paper 08-09. University of California, Los Angeles.

CBO. 2009. *The Economic Effects of Legislation to Reduce Greenhouse Gas Emissions.* Washington, DC: CBO.

CEA (Council of Economic Advisors). 2009. Economic Report of the President. Prepared by the Executive Office of the President, Washington, DC.

Chamberlain, A. 2009. *Who Pays for Climate Policy? New Estimates of the Household Burden and Economic Impact of a U.S. Cap-and-Trade System.* Washington, DC: Tax Foundation.

City of Austin. 2009. *Energy Conservation Audit and Disclosure Ordinance (ECAD).* Ordinance No. 20081106-047. Available at *http://www.austinenergy.com/about%20us/environmental%20initiatives/ordinance/index.htm.* Accessed on July 27, 2010.

City of Berkeley. 2008. *Residential Energy Conservation Ordinance (RECO).* Berkeley Municipal Code Chapter 19.16, Resolution No. 62,181–N.S. A Compliance Guide for Buyers and Sellers. Office of Energy and Sustainable Development. Available at *http://www.ci.berkeley.ca.us/uploadedFiles/Planning_and_Development/Level_3_-_Energy_and_Sustainable_Development/Residential%20Energy%20Conservation%20Ordinance%20Compliance%20Guide%202008. pdf.* Accessed on July 27, 2010.

Clarke, L., J. Edmonds, V. Krey, R. Richels, S. Rose, and M. Tavoni. 2009. International climate policy architectures: Overview of the EMF 22 International Scenarios. *Energy Economics* 31(Suppl 2):S64-S81.

Clarke, L. E. 2007. *Scenarios of Greenhouse Gas Emissions and Atmospheric Concentrations.* Washington, DC: U.S. Climate Change Science Program.

Cohen, L. R., and R. G. Noll. 1991. *The Technology Pork Barrel.* Washington, DC: Brookings Institution Press.

Committee on Climate Change. 2009. *Building a Low-Carbon Economy: The UK's Contribution to Tackling Climate Change.* London: The Stationery Office.

Conant, R. T., and K. Paustian. 2002. Spatial variability of soil organic carbon in grasslands: Implications for detecting change at different scales. *Environmental Pollution* 116(Suppl 1):S127-S135.

Cory, K. S., and B. G. Swezey. 2007. *Renewable Portfolio Standards in the States Balancing Goals and Implementation Strategies.* Golden, CO: National Renewable Energy Laboratory.

Coyle, K. 2005. *Environmental Literacy in America: What Ten Years of NEETF/Roper Research and Related Studies Say About Environmental Literacy in the U.S.* Washington, DC: National Environmental Education and Training Foundation.

Craford, M. G. 2008. High power LEDs for solid state lighting: Status, trends, and challenges. *Journal of Light and Visual Environment* 32(2):58-62.

CSPO and CATF (Consortium for Science, Policy & Outcomes and Clean Air Task Force). 2009. *Innovation Policy for Climate Change: A Report to the Nation.* Joint project of the Consortium for Science, Policy & Outcomes and Clean Air Task Force, Boston, MA.

Curry, T., S. Ansolabehere, and H. Herzog. 2007. *A Survey of Public Attitudes Towards Climate Change and Climate Change Mitigation Technologies in the United States: Analyses of 2006 Results.* Cambridge, MA: Laboratory for Energy and the Environment, Massachusetts Institute of Technology.

de la Chesnaye, F., and J. Weyant (eds). 2006. Multi-greenhouse gas mitigation and climate policy. *The Energy Journal,* Special Issue.

de la Chesnaye, F. C., C. Delhotal, B. Deangelo, D. Ottinger-Schaefer, and D. Godwin. 2007. Past, present, and future of non-CO_2 gas mitigation analysis. In *Human-Induced Climate Change: An Interdisciplinary Assessment,* M. E. Schlesinger, ed. New York: Cambridge University Press.

Dewar, M. E. 1998. Why state and local economic development programs cause so little economic development. *Economic Development Quarterly* 12(1):68-87.

Dietz, T., G. T. Gardner, J. Gilligan, P. C. Stern, and M. P. Vandenbergh. 2009. Household actions can provide a behavioral wedge to rapidly reduce U.S. carbon emissions. *Proceedings of the National Academy of Sciences of the USA* 106:18452-18456.

Dinan, T., and D. L. Rogers. 2002. Distributional effects of carbon allowance trading: How government decisions determine winners and losers. *National Tax Journal* 55(2):199-221.

Dinan, T. M. 2009. Reducing greenhouse gas emissions with a tax or a cap: Implications for efficiency and cost effectiveness. *National Tax Journal* 62:535-553.

DIUS (Department for Innovation Universities and Skills). 2009. *2008 R&D Scoreboard.* London: Department for Innovation, Universities & Skills.

DOE (Department of Energy). 2006a. *Energy Savings Potential of Solid State Lighting in General Illumination Applications.* Report prepared by Navigant Consulting for the Office of Energy Efficiency and Renewable Energy, U.S. Department of Energy, Washington, DC.

DOE. 2006b, October. *Energy Bandwidth for Petroleum Refining Processes.* Prepared by Energetics Incorporated. Washington, DC: DOE. Available at *http://www1.eere.energy.gov/industry/petroleum_refining/bandwidth.html.*

DOE. 2009. *Strategies for the Commercialization and Deployment of Greenhouse Gas Intensity-Reducing Technologies and Practices.* Washington, DC: DOE.

Downs, G. W., K. W. Danish, and P. N. Barsoom. 2000. The transformational model of international regime design: Triumph of hope or experience? *Columbia Journal of Transnational Law* 38(3):465-514.

Dudek, D. J., and J. Palmisano. 1988. Emissions trading: Why is this thoroughbred hobbled? *Columbia Journal of Environmental Law* 13(2):217-256.

Duhigg, D. 2009. EPA vows better effort on water. *New York Times.* October 15. Available at *http://www.nytimes.com/2009/10/16/business/energy-environment/16water.html?_r=1.* Accessed February 22, 2010.

Edenhofer, O., B. Knopf, T. Barker, L. Baumstark, E. Bellevrat, B. Château, P. Criqui, M. Isaac, A. Kitous, S. Kypreos, M. Leimbach, K. Lessmann, B. Magné, S. Scrieciu, H. Turton, and D. P. van Vuuren. 2010. The economics of low stabilization: Model comparison of mitigation strategies and costs. *The Energy Journal* 31(1).

EIA (Energy Information Administration). 2009. *Annual Energy Review 2008, U.S. Energy Information Administration.* Washington, DC: U.S. Government Printing Office.

EIA. 2010. *Annual Energy Outlook 2010 Early Release Overview.* Available at *http://www.eia.doe.gov/oiaf/aeo/.* Accessed February 22, 2010.

Elbakidze, L., and B. A. McCarl. 2007. Sequestration offsets versus direct emission reductions: Consideration of environmental co-effects. *Ecological Economics* 60 (3):564-571.

Ellerman, A. D. 2000. *Markets for Clean Air: The U.S. Acid Rain Program.* New York: Cambridge University Press.

Elliott, J. R., and J. Pais. 2006. Race, class, and Hurricane Katrina: Social differences in human responses to disaster. *Social Science Research* 35(2):295-321.

EPA (Environmental Protection Agency). 2006. *Global Mitigation of Non-CO2 Greenhouse Gases.* 430-R-06-005. Washington, DC: EPA.

EPA. 2009. *Inventory of U.S. Greenhouse Gas Emissions and Sinks: 1990-2007.* Washington, DC: EPA.

Fargione, J., J. Hill, D. Tilman, S. Polasky, and P. Hawthorne. 2008. Land clearing and the biofuel carbon debt. *Science* 319(5867):1235-1238.

Fawcett, A. A., K. V. Calvin, F. C. de la Chesnaye, J. M. Reilly, and J. P. Weyant. 2009. Overview of EMF 22 U.S. transition scenarios. *Energy Economics* 31(Suppl 2):S198-S211.

Fischer, C., and A. K. Fox. 2009. *Combining Rebates with Carbon Taxes: Optimal Strategies for Coping with Emissions Leakage and Tax Interactions.* Washington, DC: Resources for the Future.

Fischer, C., and R. G. Newell. 2008. Environmental and technology policies for climate mitigation. *Journal of Environmental Economics and Management* 55(2):142-162.

Fisher, B. S., N. Nakicenovic, K. Alfsen, J. C. Morlot, F. de la Chesnaye, J.-C. Hourcade, K. Jiang, M. Kainuma, E. L. Rovere, A. Matysek, A. Rana, K. Riahi, R. Richels, S. Rose, D. V. Vuuren, and R. Warren. 2007. Issues related to mitigation in the long term context. In *Climate Change 2007, Mitigation of Climate Change: Contribution of Working Group III to the Fourth Assessment Report of the Intergovernmental Panel on Climate Change,* B. Metz, O. R. Davidson, P. R. Bosch, R. Dave, and L. A. Meyer, eds. Cambridge: Cambridge University Press.

Fisher, C. 2009. *The Role of Technology Policies in Climate Mitigation.* Washington, DC: Resources for the Future.

Forster, P. and V. Ramaswamy. 2007. Changes in atmospheric constituents and radiative forcing. In *Climate Change 2007: The Physical Science Basis, Contribution of Working Group I to the Fourth Assessment Report of the IPCC.* S. Solomon, D. Qin, M. Manning, Z. Chen, M. Marquis, K. B. Averyt, M. Tignor, and H. L. Miller, eds. Cambridge: Cambridge University Press. 996 pp.

Fowlie, M., S. P. Holland, and E. T. Mansur. 2009. *What Do Emissions Markets Deliver and to Whom? Evidence from Southern California's NO$_x$ Trading Program*. National Bureau of Economic Research Working Paper Series 15082.

Frankel, J. 2009. A proposal for specific formulas and emissions targets for all countries in all decades. In *The Competitiveness Impacts of Climate Change Mitigation Policies*, J. E. Aldy and W. A. Pizer, eds. Arlington, VA: Pew Center on Global Climate Change.

Galitsky, C., S. Chang, E. Worrell, and E. Masanet. 2005. *Energy Efficiency Improvement and Cost Saving Opportunities for Petroleum Refineries, An ENERGY STAR Guide for Energy and Plant Managers*. LBNL-57260-Revision. Berkeley, CA: Lawrence Berkeley National Laboratory.

Gallagher, K. S., and L. D. Anadon. 2009. *DOE Budget Authority for Energy Research, Development, and Demonstration Database*. Energy Technology Innovation Policy, John F. Kennedy School of Government, Harvard University.

GAO (Government Accountability Office). 1996. Federal research: Changes in electricity-related R&D funding. In *Report to the Ranking Minority Member, Committee on Science, House of Representatives*. Washington, DC: GAO.

GAO. 2007. *Long-standing Problems with DOE's Program for Setting Efficiency Standards Continue to Result in Foregone Energy Savings*. Washington, DC: GAO.

Gardner, G. T., and P. C. Stern. 2002. *Environmental Problems and Human Behavior*, 2nd ed. Boston, MA: Pearson Custom Publishing.

Gardner, G. T., and P. C. Stern. 2008. The most effective actions U.S. households can take to curb climate change. *Environment* 50(5):12-24.

German Advisory Council on Global Change. 2009. *Solving the Climate Dilemma: The Budget Approach*. Available at *http:// www.wbgu.de/wbgu_sn2009_en.html*. Accessed February 15, 2010.

Gillingham, K., R. G. Newell, and K. Palmer. 2009. *Energy Efficiency Economics and Policy*. Washington, DC. Resources for the Future.

Goettle, R. J., and A. A. Fawcett. 2009. The structural effects of cap and trade climate policy. *Energy Economics* 31(2): S63-S306.

Gold, R., L. Furrey, S. Nadel, J. S. Laitner, and R. N. Elliott. 2009. *Energy Efficiency in the American Clean Energy and Security Act of 2009: Impacts of Current Provisions and Opportunities to Enhance the Legislation*. Washington, DC: The American Council for an Energy-Efficient Economy.

Golombek, R., and M. Hoel. 2009. *International Cooperation on Climate-Friendly Technologies*. CESifo Working Paper Series 2677. Social Science Research Network. Available at *http://ssrn.com/abstract=1427132*.

Goulder, L. H. 1997. Environmental taxation in a second-best world. Pp. 28-54 in *The International Yearbook of Environmental and Resource Economics 1997/1998*, T. Tietenberg and H. Folmer, eds. Cheltenham, U.K.: Edward Elgar.

Goulder, L. H., and I. W. H. Parry. 2008. Instrument choice in environmental policy. *Review of Environmental Economics and Policy* 2(2):152-174.

Goulder, L. H., M. A. Hafstead, and M. Dworsky. 2009. *Impacts of Alternative Emissions Allowance Allocation Methods Under a Federal Cap-and-Trade Program*. Cambridge, MA: National Bureau of Economic Research.

Grainger, C. A., and C. D. Kolstad. 2009. *Who Pays a Price on Carbon?* Cambridge, MA: National Bureau of Economic Research.

Granade, H. C. 2009. *Unlocking Energy Efficiency in the U.S. Economy*. Washington, DC: McKinsey.

Grimmett, J. J., and L. Parker. 2008. *Whether Import Requirements Contained in Title VI of S. 2191, the Lieberman-Warner Climate Security Act of 2008, as Ordered Reported, Are Consistent with U.S. WTO Obligations*. Washington, DC: Congressional Research Service.

Grübler, A. 2004. *Transitions in Energy Use*. Laxenburg, Austria: International Institute for Applied Systems Analysis.

Hahn, R. W., and G. L. Hester. 1989. Where did all the markets go? An analysis of EPA's Emission Trading Program. *Yale Journal of Regulation* 6(1):109-153.

Hall, D. S., M. Levi, W. A. Pizer, and T. Ueno. 2008. *Policies for Developing Country Engagement*. Discussion Paper 08-15, Harvard Project on International Climate Agreements. Cambridge, MA.

Hamin, E. M., and N. Gurran. 2009. Urban form and climate change: Balancing adaptation and mitigation in the U.S. and Australia. *Habitat International* 33(3):238-245.

Hanemann, M. 2009. The role of emission trading in domestic climate policy. *The Energy Journal* 30(Special Issue 2).

Hansen, J., M. Sato, P. Kharecha, D. Beerling, R. Berner, V. Masson-Delmotte, M. Pagani, M. Raymo, D. L. Royer, and J. C. Zachos. 2008. Target atmospheric CO_2: Where should humanity aim? *Open Atmospheric Science Journal* 2:217-231.

Hawkins, D. G., D. A. Lake, D. L. Nielson, and M. J. Tierney. 2006. *Delegation and Agency in International Organizations*. New York: Cambridge University Press.

Hirst, E. 1988. The Hood River Conservation Project—An evaluator's dream. *Evaluation Review* 12(3):310-325.

Hirst, E., and F. M. O'Hara. 1986. *Energy Efficiency in Buildings: Progress and Promise, Series on Energy Conservation and Energy Policy*. Washington, DC: American Council for an Energy-Efficient Economy.

Hoel, M., and L. Karp. 2001. Taxes and quotas for a stock pollutant with multiplicative uncertainty. *Journal of Public Economics* 82(1):91-114.

Hoel, M., and L. Karp. 2002. Taxes versus quotas for a stock pollutant. *Resource and Energy Economics* 24:367-384.

Hoerner, J. A., and N. Robinson. 2008. *A Climate of Change. African Americans, Global Warming, and a Just Climate Policy for the U.S.* Oakland, CA: Environmental Justice and Climate Change.

Hoffman, A. J., and R. Henn. 2008. Overcoming the social and psychological barriers to green building. *Organization and Environment* 21(4):390-419.

Howitt, A. M., and A. Altshuler. 1999. The politics of controlling auto air pollution. In *Essays in Transportation Economics and Policy*. Washington, DC: The Brookings Institution.

Huntington, H. G. 2008. The oil security problem. In *International Handbook on the Economics of Energy*, L. C. Hunt and J. Evans, eds. Cheltenham, U.K.: Edward Elgar.

IEA (International Energy Agency). 2004. *Oil Crises and Climate Challenges: 30 Years of Energy Use in IEA Countries*. Paris: Organisation for Economic Co-operation and Development.

IEA. 2008. *World Energy Outlook 2008*. Paris: International Energy Agency.

IEA. 2009a. *IEA Technology Roadmaps*. Available at *http://www.iea.org/subjectqueries/keyresult.asp?KEYWORD_ID=4156*. Accessed January 27, 2010.

IEA. 2009b. *Ensuring Green Growth in a Time of Economic Crisis: The Role of Energy Technology*. Paris: IEA, p. 7.

Ikenberry, G. J. 2000. *After Victory*. Princeton, NJ: Princeton University Press.

IPCC (Intergovernmental Panel on Climate Change). 1995. *Report of Working Group I—the Science of Climate Change, with a Summary for Policymakers (SPM)*. J. T. Houghton, L. G. Meira Filho, B. A. Callender, N. Harris, A. Kattenberg, and K. Maskell, eds. Cambridge, U.K.: Cambridge University Press, 572 pp.

IPCC, 2001. *Climate Change 2001. Mitigation: Contribution of Working Group III to the Third Assessment Report of the Intergovernmental Panel on Climate Change*. B. Metz, O. Davidson, R. Swart and J. Pan, eds. Cambridge, U.K.: Cambridge University Press, 700 pp.

IPCC/TEAP. 2005 - Metz, B., K. Lambert, S. Solomon, S.O. Andersen, O. Davidson, J. Pons, D. de Jager, T. Kestin, M. Manning, and L. Meyer (Eds). *Safeguarding the Ozone Layer and the Global Climate System: Issues Related to Hydrofluorocarbons and Perfluorocarbons*. Cambridge, U.K.: Cambridge University Press, UK. pp 478..

IPCC. 2007a. *Climate Change 2007: Mitigation, Contribution of Working Group III to the Fourth Assessment Report of the Intergovernmental Panel on Climate Change*. B. Metz, O. R. Davidson, P. R. Bosch, R. Dave, and L. A. Meyer, eds. Cambridge, U.K.: Cambridge University Press, 851 pp.

IPCC. 2007b. *Climate Change 2007: The Physical Science Basis, Contribution of Working Group I to the Fourth Assessment Report of the IPCC*. S. Solomon, D. Qin, M. Manning, Z. Chen, M. Marquis, K. B. Averyt, M. Tignor, and H. L. Miller, eds. Cambridge, U.K.: Cambridge University Press. 996 pp.

Jacoby, H. D., and A. D. Ellerman. 2004. The safety valve and climate policy. *Energy Policy* 32(4):481-491.

Jaffe, J., and R. N. Stavins. 2009. Linkage of tradable permit systems in international climate policy architecture. In *The Competitiveness Impacts of Climate Change Mitigation Policies*, J. E. Aldy and W. A. Pizer, eds. Arlington, VA: Pew Center on Global Climate Change.

Jaffe, J. L., and R. N. Stavins. 2008. *Linkage of Tradable Permit Systems in International Climate Policy Architecture (October 16, 2008)*. FEEM Working Paper 90.2008, HKS Working Paper RWP08-053. Social Science Research Network. Available at *http://ssrn.com/abstract=1285606*.

Kahneman, D., and A. Tversky. 1979. Prospect theory: Analysis of decision under risk *Econometrica* 47(2):263-291.

Keith, D. W., M. Ha-Duong, and J. K. Stolaroff. 2006. Climate strategy with CO_2 capture from the air. *Climatic Change* 74(1-3):17-45.

Keohane, N. O. 2009. Cap and trade, rehabilitated: Using tradable permits to control U.S. greenhouse gases. *Review of Environmental Economics and Policy* 3(1):42-62.

Keohane, R. O. 1984. *After Hegemony: Cooperation and Discord in the World Political Economy.* Ewing, NJ: Princeton University Press.

Keohane, R. O., and K. Raustiala. 2009. Toward a post-Kyoto climate change architecture: A political analysis. In *The Competitiveness Impacts of Climate Change Mitigation Policies,* J. E. Aldy and W. A. Pizer, eds. Arlington, VA: Pew Center on Global Climate Change.

Kim, M. 2004. *Economic Investigation of Discount Factors for Agricultural Greenhouse Gas Emission Offsets.* Department of Agricultural Economics, Texas A&M University, College Station, TX.

Kindermann, G., M. Obersteiner, B. Sohngen, J. Sathaye, K. Andrasko, E. Rametsteiner, B. Schlamadinger, S. Wunder, and R. Beach. 2008. Global cost estimates of reducing carbon emissions through avoided deforestation. *Proceedings of the National Academy of Sciences of the USA* 105(30):10302-10307.

Kingsbury, B., R. B. Stewart, and B. Rudyk. 2009. *Climate Financing: Regulatory and Funding Strategies for Climate Change and Global Development.* New York: NYU Press.

Knittel, C. R. 2009. *The Implied Cost of Carbon Dioxide Under the Cash for Clunkers Program,* CSEM Working Paper 189, University of California Energy Institute, Center for the Study of Energy Markets, August.

Kovats, R. S., and S. Hajat. 2008. Heat stress and public health: A critical review. *Annual Review of Public Health* 29:41-55.

Krey, V., and K. Riahi. 2009. Implications of delayed participation and technology failure for the feasibility, costs, and likelihood of staying below temperature targets: Greenhouse gas mitigation scenarios for the 21st century. *Energy Economics* 31(Suppl 2):S94-S106.

Kyle, P., L. Clarke, G. Pugh, M. Wise, K. Calvin, J. Edmonds, and S. Kim. 2009. The value of advanced technology in meeting 2050 greenhouse gas emissions targets in the United States. *Energy Economics* 31(2):S254-S267.

Lackner, K., H.-J. Ziock, and P. Grimes. 1999. Carbon dioxide extraction from air: Is it an option? Paper presented at the 24th International Conference on Coal Utilization & Fuel Systems, Clearwater, FL.

Lazarus, R. J. 2004. *The Making of Environmental Law.* Chicago, IL: University of Chicago Press.

Lee, J., F. M. Veloso, D. A. Hounshell, and E. S. Rubin. 2010. Forcing technological change: A case of automobile emissions control technology development in the US. *Technovation* 30:249-264.

Lehmann, J., J. Gaunt, and M. Rondon. 2006. Bio-char sequestration in terrestrial ecosystems: A review. *Mitigation and Adaptation Strategies for Global Change* 11(2):403-427.

Leiby, P. N. 2007. *Estimating the Energy Security Benefits of Reduced U.S. Oil Imports.* Oak Ridge, TN: Oak Ridge National Laboratory.

Leiserowitz, A., E. Maibach, and C. Roser-Renouf. 2008. *Global Warming's Six Americas: An Audience Segmentation.* New Haven, CT: Yale University. Available at *http://research.yale.edu/environment/climate.* Accessed January 27, 2010.

London Convention. 1972. *Convention on the Prevention of Marine Pollution by Dumping of Wastes and Other Matter.* Available at *http://www.imo.org/home.asp?topic_id=1488.* Accessed February 16, 2010.

Lutsey, N., and D. Sperling. 2008. America's bottom-up climate change mitigation policy. *Energy Policy* 36(2):673-685.

Lutzenhiser, L., L. Cesafsky, H. Chappells, M. Gossard, M. Moezzi, D. Moran, J. Peters, M. Spahic, P. Stern, E. Simmons, and H. Wilhite. 2009. *Behavioral Assumptions Underlying California Residential Sector Energy Efficiency Programs.* Portland State University, Center for Urban Studies, Portland, OR. Report to the California Institute for Energy and Environment and the California Public Utilities Commission, Oakland, CA.

Marland, G., B. A. McCarl, and U. Schneider. 2001. Soil carbon: Policy and economics. *Climatic Change* 51(1):101-117.

Martin, L. L. 1992. Interests, power, and multilateralism. *International Organization* 46(4):765-792.

McCarl, B. A., and J. M. Reilly. 2007. *Agriculture in the Climate Change and Energy Price Squeeze: Part 2: Mitigation Opportunities.* Department of Agricultural Economics, Texas A&M University.

McCarl, B. A., and U. A. Schneider. 2000. U.S. agriculure's role in a greenhouse gas emission mitigation world: An economic perspective. *Review of Agricultural Economics* 22(1):134-159.

McCarl, B. A., and U. A. Schneider. 2001. Climate change: Greenhouse gas mitigation in U.S. agriculture and forestry. *Science* 294(5551):2481-2482.

McCarl, B. A., C. Peacocke, R. Chrisman, C.-C. Kung, and R. D. Sands. 2009. Economics of biochar production, utilisation and GHG offsets. In *Biochar for Environmental Management: Science and Technology*, J. Lehmann and S. Joseph, eds. London: Earthscan Productions.

McGuinness, M., and A. D. Ellerman. 2008. *The Effects of Interactions Between Federal and State Climate Policies*. Cambridge, MA: MIT Center for Energy and Environmental Policy Research.

McHenry, M. P. 2009. Agricultural bio-char production, renewable energy generation and farm carbon sequestration in Western Australia: Certainty, uncertainty and risk. *Agriculture, Ecosystems and Environment* 129(1-3):1-7.

Meehl, G. A., and T. F. Stocker. 2007. Global climate projections. In *Climate Change 2007: The Physical Science Basis, Contribution of Working Group I to the Fourth Assessment Report of the IPCC*, S. Solomon, D. Qin, M. Manning, Z. Chen, M. Marquis, K. B. Averyt, M. Tignor, and H. L. Miller, eds. Cambridge, U.K.: Cambridge University Press, 996 pp.

Melillo, J. M., J. M. Reilly, D. W. Kicklighter, A. C. Gurgel, T. W. Cronin, S. Paltsev, B. S. Felzer, X. Wang, A. P. Sokolov, and C. A. Schlosser. 2009. Indirect emissions from biofuels: How important? *Science* 326(5958):1397-1399.

Metcalf, G. E. 2009. Designing a carbon tax to reduce U.S. greenhouse gas emissions. *Review of Environmental Economics and Policy* 3 (1):63-83.

METI (Ministry of Economy, Trade, and Industry). 2008. *Top Runner Program: Developing the World's Best Energy-Efficient Appliances*. Tokyo: METI.

Metz, B., L. Kuijpers, S. Solomon, S. O. Andersen, O. Davidson, J. Pons, D. de Jager, T. Kestin, M. Manning, and L. Meyer, eds. 2005. *Safeguarding the Ozone Layer and the Global Climate System: Issues Related to Hydrofluorocarbons and Perfluorocarbons*. Cambridge, U.K.: Cambridge University Press, 478 pp.

Michaelowa, A., and F. Jotzo. 2005. Transaction costs, institutional rigidities and the size of the clean development mechanism. *Energy Policy* 33(4):511-523.

Mooney, S., J. Antle, S. Capalbo, and K. Paustian. 2004. Design and costs of a measurement protocol for trades in soil carbon credits. *Canadian Journal of Agricultural Economics* 52(3):257-287.

Morello-Frosch, R., M. Pastor, J. Sadd, and S. B. Shonkoff. 2009. *The Climate Gap*. Los Angeles, CA: Pere Publications.

Morris, J. 2009. *Combining a Renewable Portfolio Standard with a Cap-and-Trade Policy: A General Equilibrium Analysis*. Cambridge: Massachusetts Institute of Technology.

Mueller, S. 2006. Missing the spark: An investigation into the low adoption paradox of combined heat and power technologies. *Energy Policy* 34:3153-3164.

Murray, B. C., B. A. McCarl, and H. C. Lee. 2004. Estimating leakage from forest carbon sequestration programs. *Land Economics* 80(1):109-124.

Murray, B. C., A. J. Sommer, B. Depro, B. L. Sohngen, B. A. Mccarl, D. Gillig, B. De Angelo, and K. Andrasko. 2005. *Greenhouse Gas Mitigation Potential in US Forestry and Agriculture*. Washington, DC: Environmental Protection Agency.

Murray, B. C., R. G. Newell, and W. A. Pizer. 2009a. Balancing cost and emissions certainty: An allowance reserve for cap-and-trade. *Review of Environmental Economics and Policy* 3(1):84-103.

Murray, B. C., R. N. Lubowski, and B. L. Sohngen. 2009b. *Including International Forest Carbon Incentives in Climate Policy: Understanding the Economics*. Nicholas Institute Report. Durham, NC: Nicholas Institute for Environmental Policy Solutions, Duke University.

Myers, D., and E. Gearin. 2001. Current preferences and future demand for denser residential environments. *Housing Policy Debate* 12(4):633-659.

Nakano, S., A. Okamura, N. Sakurai, M. Suzuki, Y. Tojo, and N. Yamano. 2009. *The Measurement of CO_2 Embodiments in International Trade: Evidence from the Harmonised Input-Output and Bilateral Trade Database*. Organization for Economic Co-operation and Development (OECD) Science, Technology and Industry Working Papers 2009/3, OECD, Directorate for Science, Technology, and Industry.

Nakicenovic, N., et al. 2000. *Special Report on Emissions Scenarios: A Special Report of Working Group III of the Intergovernmental Panel on Climate Change*. Cambridge, U.K.: Cambridge University Press, 599 pp. Available at *http://www.grida.no/climate/ipcc/emission/index.htm*.

Newell, R. G. 2009. International climate technology strategies. In *The Competitiveness Impacts of Climate Change Mitigation Policies*, J. E. Aldy and W. A. Pizer, eds. Arlington, VA: Pew Center on Global Climate Change.

Newell, R. G., and W. A. Pizer. 2003. Discounting the distant future: How much do uncertain rates increase valuations? *Journal of Environmental Economics and Management* 46(1):52-71.

Nordhaus, W. D. 2008. *A Question of Balance: Weighing the Options on Global Warming Policies*. New Haven, CT: Yale University Press.

Nordqvist, J. 2006. Evaluation of Japan's Top Runner Programme. Energy Intelligence for Europe Program.

Noyelle, T. J. 1987. *Beyond Industrial Dualism: Market and Job Segmentation in the New Economy*. Boulder, CO: Westview Press.

NRC (National Research Council). 1999. *Sharing the Fish: Toward a National Policy on Individual Fishing Quotas*. Washington, DC: National Academy Press.

NRC. 2001. *Energy Research at DOE: Was It Worth It?* Washington, DC: National Academies Press.

NRC. 2002a. *New Tools for Environmental Protection: Education, Information, and Voluntary Measures*. Washington, DC: National Academies Press.

NRC. 2002b. *Effectiveness and Impact of Corporate Average Fuel Economy (CAFE) Standards*. Washington, DC: National Academies Press.

NRC. 2005. *Decision Making for the Environment: Social and Behavioral Science Research Priorities*. Washington, DC: National Academies Press.

NRC. 2008. *Public Participation in Environmental Assessment and Decision Making*. Washington, DC: National Academies Press.

NRC. 2009a. *America's Energy Future: Technology and Transformation*. Washington, DC: National Academies Press.

NRC. 2009b. *America's Energy Future: Real Prospects for Energy Efficiency in the United States*. Washington, DC: National Academies Press.

NRC. 2009c. *Hidden Costs of Energy: Unpriced Consequences of Energy Production and Use*. Washington, DC: National Academies Press.

NRC. 2009d. *TRB Special Report 298: Driving and the Built Environment: Effects of Compact Development on Motorized Travel, Energy Use, and CO_2 Emissions*. Washington, DC: National Academies Press.

NRC. 2010a. *ACC: Advancing the Science of Climate Change*. Washington, DC. National Academies Press.

NRC. 2010b. *ACC: Informing an Effective Response to Climate Change*. Washington, DC: National Academies Press.

NRC. 2010c. *ACC: Adapting to the Impacts of Climate Change*. Washington, DC: National Academies Press.

NSB (National Science Board). 2009. *Building a Sustainable Energy Future: U.S. Actions for an Effective Energy Economy Transformation*. Washington, DC: National Science Foundation.

NSF (National Science Foundation). 2008. *Science and Engineering Indicators*. Arlington, VA: National Science Board.

Ostrom, E. 1990. *Governing the Commons: The Evolution of Institutions for Collective Action*. New York: Cambridge University Press.

Ostrom, E. 2002. Reformulating the Commons. *Ambiente & Sociedade* 5(10).

Ostrom, E. 2010. A multi-scale approach to coping with climate change and other collective action problems. *Solutions Journal* 2(2010).

Ottmar, E. , B. Knopf, T. Barker, L. Baumstark, E. Bellevrat, B. Château, P. Criqui, M. Isaac, A. Kitous, S. Kypreos, M. Leimbach, K. Lessmann, B. Magné, S. Scrieciu, H. Turton and D. P. van Vuuren. 2010. The Economics of Low Stabilization: Model Comparison of Mitigation Strategies and Costs. *The Energy Journal, International Association for Energy Economics* 0(Special I): 11-48.

Oye, K. N., and J. H. Maxwell. 1995. Local commons and global interdependence. In *Local Commons and Global Interdependence: Heterogeneity and Cooperation in Two Domains*, R. O. Keohane and E. Ostrom, eds. London: Sage.

Palmer, M. A., E. S. Bernhardt, W. H. Schlesinger, K. N. Eshleman, E. Foufoula-Georgiou, M. S. Hendryx, A. D. Lemly, G. E. Likens, O. L. Loucks, M. E. Power, P. S. White, and P. R. Wilcock. 2010. Mountaintop mining consequences. *Science* 327(5962):148-149.

Paltsev, S., J. M. Reilly, H. D. Jacoby, A. C. Gurgel, G. E. Metcalf, A. P. Sokolov, and J. F. Holak. 2007. *Assessment of U.S. Cap-and-Trade Proposals*. Cambridge, MA: National Bureau of Economic Research.

Paltsev, S., J. Reilly, H. Jacoby, and J. Morris. 2009. The cost of climate policy in the United States. *Energy Economics* 3(2): S235-S243.

Parry, I. W. H., H. Sigman, M. Walls, and R. C. Williams. 2006. The incidence of pollution control policies. *The International Yearbook of Environmental and Resource Economics 2006/2007*:1-42.

Parry, I., W. Walls, and W. Harrington. 2007. *Automobile Externalities and Policies*. Washington, DC: Resources for the Future.

Parson, E. 2003. *Protecting the Ozone Layer: Science and Strategy*. Oxford: Oxford University Press.

Pastor, M., Jr. 2007. Quién es más urbanista? Latinos and smart growth. Pp. 73-102 in *Growing Smarter: Achieving Livable Communities, Environmental Justice, and Regional Equity*, R. D. Bullard, ed. Cambridge, MA: MIT Press.

Patashnik, E. M. 2008. *Reforms at Risk: What Happens After Major Policy Changes are Enacted*. Princeton, NJ: Princeton University Press.

Paustian, K., J. Brenner, M. Easter, K. Killian, S. Ogle, C. Olson, J. Schuler, R. Vining, and S. Williams. 2009. Counting carbon on the farm: Reaping the benefits of carbon offset programs. *Journal of Soil and Water Conservation* 64(1):36A-40A.

PCAST (President's Council of Advisors on Science and Technology). 2008. *The Energy Imperative: Report Update*. Washington, DC: PCAST.

Pearce, D. 2003. *Conceptual Framework for Analyzing the Distributive Impacts of Environmental Policies*. Prepared for the OECD Environment Directorate Workshop on the Distribution of Benefits and Costs of Environmental Policies, Paris. Available at *http://www.ucl.ac.uk/~uctpa36/oecd%20distribution.pdf*.

Peters, A. H., and P. S. Fisher. 2002. *State Enterprise Zone Programs: Have They Worked?* Kalamazoo, MI: W.E. Upjohn Institute for Employment Research.

Pew Center. 2007, May. *International Sectoral Agreements in a Post-2012 Climate Framework*. Prepared for the Pew Center on Global Climate Change. Daniel Bodansky. Available at *http://www.pewclimate.org/working-papers/sectoral*. Accessed February 16, 2010.

Pew Center. 2009a. *Fewer Americans See Solid Evidence of Global Warming. Modest Support for "Cap and Trade" Policy*. Washington, DC: Pew Research Center for the People and the Press. Available at *http://people-press.org/reports/pdf/556.pdf*. Accessed February 16, 2010.

Pew Center. 2009b. *The Competitiveness Impacts of Climate Change Mitigation Policies*. Available at *http://www.pewclimate.org/international/CompetitivenessImpacts*. Accessed February 16, 2010.

Pew Center. 2009c. *The Clean Energy Economy: Repowering Jobs, Business, and Investments Across America*. Available at *http://www.pewcenteronthestates.org/uploadedFiles/Clean_Economy_Report_Web.pdf*. Accessed February 16, 2010.

Pizer, W. A. 2002. Combining price and quantity controls to mitigate global climate change. *Journal of Public Economics* 85(3):409-434.

Pollak, M. F., and E. J. Wilson. 2009. Regulating geologic sequestration in the US: Early rules take divergent approaches. *Environmental Science and Technology* 43(9):3035-3041.

Pollin, R., H. Garrett-Peltier, J. Heintz, and H. Scharber, 2008. *Green Recovery: A Program to Create Good Jobs and Start Building a Low-Carbon Economy*. Center for American Progress and Political Economy Research Institute: University of Massachusetts, Amherst.

Popp, D. C., R. G. Newell, and A. B. Jaffe. 2009. *Energy, the Environment, and Technological Change*. Cambridge, MA: National Bureau of Economic Research.

Post, W. M., J. E. Amonette, R. Birdsey, C. T. Garten Jr., R. C. Izaurralde, P. M. Jardine, J. Jastrow, R. Lal, G. Marland, B. A. McCarl, A. M. Thomson, T. O. West, S. D. Wullschleger, and F. B. Metting. 2009. Terrestrial biological carbon sequestration: Science for enhancement and implementation. In *Science and Technology of Carbon Sequestration*, B. Mcpherson and E. Sundquist, eds. Washington, DC: American Geophysical Union.

Prindle, B. 2007. *Quantifying the Effects of Market Failures in the End-Use of Energy*. Washington, DC: American Council for an Energy-Efficient Economy.

Putnam, R. D. 1988. Diplomacy and domestic politics: The logic of two-level games. *International Organization* 42(3):427-460.

Qiu, L. D., and Z. Tao. 1998. Policy on international R&D cooperation: Subsidy or tax? *European Economic Review* 42(9):1727-1750.

Rai, V., and D. G. Victor. 2009. Climate change and the energy challenge: A pragmatic approach for India. *Economic and Political Weekly* 44(31):78-85.

Ramaswamy, V., O. Boucher, J. Haigh, D. Hauglustine, J. Haywood, G. Myhre, T. Nakajima, G. Y. Shi, and S. Solomon. 2001. Radiative forcing of climate. Pp. 349-416 in *Climate Change 2001: The Scientific Basis. Contribution of Working Group I to the Third Assessment Report of the Intergovernmental Panel on Climate Change.* Cambridge, U.K.: Cambridge University Press.

Reilly, J., R. Prinn, J. Harnisch, J. Fitzmaurice, H. Jacoby, D. Kicklighter, J. Melillo, P. Stone, A. Sokolov, and C. Wang. 1999. Multi-gas assessment of the Kyoto protocol. *Nature* 401(6753):549-555.

Revesz, R. L. 2001. Federalism and environmental regulation: A public choice analysis. *Harvard Law Review* 115(2):553-641.

Rosa, E. A., and R. L. Clark, Jr. 1999. Historical routes to technological gridlock: Nuclear technology as prototypical vehicle. *Research in Social Problems and Public Policy* 7:21-57.

Rosa, E. A., and R. E. Dunlap. 1994. Nuclear power: Three decades of public opinion. *Public Opinion Quarterly* 58:295-325.

Rose, A., and G. Oladosu. 2002. Greenhouse gas reduction policy in the United States: Identifying winners and losers in an expanded permit trading system. *The Energy Journal* 23(1):1-18.

Rosenfeld, A. H., and H. Akbari. 2008. *White Roofs Cool the World, Directly Offset CO_2 and Delay Global Warming: Research Highlights.* Sacramento, CA: California Energy Commission.

Rubin, E. S. 2005. *The Government Role in Technology Innovation: Lessons for the Climate Change Policy Agenda.* Presented at the 10th Biennial Conference on Transportation Energy and Environmental Policy. Institute of Transportation Studies, University of California, Davis.

Ruderman, H., M. D. Levine, and J. E. Mcmahon. 1987. The behavior of the market for energy efficiency in residential appliances including heating and cooling. *The Energy Journal* 8:101-124.

Samaras, C., J. Apt, I. L. Azevedo, L. Lave, G. Morgan, and E. S. Rubin. 2009. *Cap and Trade is Not Enough: Improving U. S. Climate Policy: A Briefing Note from the Department of Engineering and Public Policy.* Washington, DC: Carnegie Mellon University.

Schafer, A., J. B. Heywood, H. D. Jacoby, and I. A. Waitz. 2009. *Transportation in a Climate-Constrained World.* Cambridge, MA: MIT Press.

Schelling, T. C. 1960. *The Strategy of Conflict.* Cambridge, MA: Harvard University Press.

Schweitzer, L., and M. Stephenson. 2007. Right answers, wrong questions: Environmental justice as urban research. *Urban Studies* 44(2):319-337.

Searchinger, T., R. Heimlich, R. A. Houghton, F. Dong, A. Elobeid, J. Fabiosa, S. Tokgoz, D. Hayes, and T. H. Yu. 2008. Use of U.S. croplands for biofuels increases greenhouse gases through emissions from land-use change. *Science* 319(5867):1238-1240.

Shadbegian, R. J., W. Gray, and C. L. Morgan. 2005. *Benefits and Costs from Sulfur Dioxide Trading: A Distributional Analysis.* Washington, DC: Environmental Protection Agency.

Shammin, M. R., and C. W. Bullard. 2009. Impact of cap-and-trade policies for reducing greenhouse gas emissions on U.S. households. *Ecological Economics* 68(8-9):2432-2438.

Sieg, H., V. K. Smith, H. S. Banzhaf, and R. Walsh. 2004. Estimating the general equilibrium benefits of large changes in spatially delineated public goods. *International Economic Review* 45(4):1047-1077.

Sijm, J., K. Neuhoff, and Y. Chen. 2006. CO_2 cost pass-through and windfall profits in the power sector. *Climate Policy* 6(1):49-72.

Small, K. A., and K. Van Dender. 2007. Fuel efficiency and motor vehicle travel: The declining rebound effect. *Energy Journal* 28(1):25-51.

Smith, G. A., B. A. McCarl, C. S. Li, J. H. Reynolds, R. Hammerschlag, R. L. Sass, W. J. Parton, S. M. Ogle, K. Paustian, J. A. Holtkamp, and W. Barbour. 2007. *Harnessing Farms and Forests in the Low-Carbon Economy: How to Create, Measure, and Verify Greenhouse Gas Offsets.* Z. Willey and W. L. Chameides, eds. Raleigh, NC: Duke University Press.

Smith, P. 2004. How long before a change in soil organic carbon can be detected? *Global Change Biology* 10(11):1878-1883.

Socolow, R. H., and A. Glaser. 2009. Balancing risks: Nuclear energy and climate change. *Daedalus* 138(4):31-44.

Sohngen, B., R. H. Beach, and K. Andrasko. 2008. Avoided deforestation as a greenhouse gas mitigation tool: Economic issues. *Journal on Environmental Quality* 37(4):1368-1375.

Sokolov, A. P., P. H. Stone, C. E. Forest, R. Prinn, M. C. Sarofim, M. Webster, S. Paltsev, C. A. Schlosser, D. Kicklighter, S. Dutkiewicz, J. Reilly, C. Wang, B. Felzer, J. M. Melillo, and H. D. Jacoby. 2009. Probabilistic forecast for twenty-first-century climate based on uncertainties in emissions (without policy) and climate parameters. *Journal of Climate* 22(19):5175-5204.

Solomon, S., G.-K. Plattner, R. Knutti, and P. Friedlingstein. 2009. Irreversible climate change due to carbon dioxide emissions. *Proceedings of the National Academy of Sciences* 106(6):1704-1709.

Sovacool, B., and R. Hirsh. 2007. Energy myth The barriers to new and innovative energy technologies are primarily technical: The case of distributed generation. Pp. 145-169 in *Energy and American Society—Thirteen Myths*, B. K. Sovacool and M. A. Brown, eds. New York: Springer.

Sovacool, B. K., and M. A. Brown. 2009. Scaling the policy response to climate change. *Policy and Society* 27(4):317-328.

Sperling, D., and D. Gordon. 2009. *Two Billion Cars: Driving Toward Sustainability*. New York: Oxford University Press.

Stephens, J. C., and D. W. Keith. 2008. Assessing geochemical carbon management. *Climatic Change* 90(3):217-242.

Stern, N. H. 2009. *The Global Deal: Climate Change and the Creation of a New Era of Progress and Prosperity*, 1st ed. New York: Public Affairs.

Stern, P. 2002. Changing behavior in households and communities: What have we learned? In *New Tools for Environmental Protection: Education, Information, and Voluntary Measures*, National Research Council, T. Dietz and P. C. Stern, eds. Washington, DC: National Academy Press.

Stern, P. C. 1986. Blind spots in policy analysis: What economics doesn't say about energy use. *Journal of Policy Analysis and Management* 5:200-227.

Stern, P. C. 2008. Environmentally significant behavior in the home. In *The Cambridge Handbook of Psychology and Economic Behavior*. A. Lewis, ed. Cambridge, U.K.: Cambridge University Press.

Stern, P. C., E. Aronson, J. M. Darley, D. H. Hill, E. Hirst, W. Kempton, and T. J. Wilbanks. 1986. The effectiveness of incentives for residential energy conservation. *Evaluation Review* 10(2):147-176.

Taylor, M. R., E. S. Rubin, and D. A. Hounshell. 2005. Control of SO_2 emissions from power plants: A case of induced technological innovation in the U.S. *Technological Forecasting and Social Change* 72(6):697-718.

Tietenberg, T. 2009. Reflections—Energy efficiency policy: Pipe dream or pipeline to the future? *Review of Environmental Economics and Policy* 3(2):304-320.

Tietenberg, T. H. 2006. *Emissions Trading: Principles and Practice*. Washington, DC: Resources for the Future.

Tran, C. I. 2006. *Equilibrium Welfare Impacts of the 1990 Clean Air Act Amendments in the Los Angeles Area*. Department of Agricultural and Resource Economics, University of Maryland.

UNEP and WTO (United Nations Environment Programme and World Trade Organization). 2009. *Trade and Climate Change*. Geneva: WTO Publications.

UNFCCC (United Nations Framework Convention on Climate Change). 2008. Analysis of Possible Means to Reach Emission Reduction Targets and of Relevant Methodological Issues; FCCC/TP/2008/2. Geneva.

U.S. Census Bureau. 2000. *Population Estimates Program, Population Division, U.S. Census Bureau*. Revised June 28, 2000. Available at *http://www.census.gov/popest/archives/1990s/popclockest.txt*. Accessed March 3, 2010.

U.S. Census Bureau. 2007. *American Factfinder*. Available at *http://factfinder.census.gov/servlet/GCTSelectedDatasetPageServlet?_lang=en&_ts=285702889855*. Accessed March 3, 2010.

Velders, G. J. M., S. O. Andersen, J. S. Daniel, D. W. Fahey, and M. Mcfarland. 2007. The importance of the Montreal Protocol in protecting climate. *Proceedings of the National Academy of Sciences of the United States of America* 104(12):4814-4819.

Velders, G. J. M., D. W. Fahey, J. S. Daniel, M. McFarland, and S. O. Andersen. 2009. The large contribution of projected HFC emissions to future climate forcing. *Proceedings of the National Academy of Sciences of the United States of America* 106(27):10949-10954.

Waltz, K. N. 1979. *Theory of International Politics*. Addison-Wesley Series in Political Science. Reading, MA: Addison-Wesley.

Wang, M. Q., and Z. Haq. 2008. Response to the article by Searchinger et al. in the February 7, 2008, Sciencexpress, "Use of U.S. Croplands for Biofuels Increases Greenhouse Gases through Emissions from Land Use Change" Letter to Science. *http://www.sciencemag.org/cgi/eletters/1151861v1*

Wara, M. 2007. Is the global carbon market working? *Nature* 445(7128):595-596.

Wassmer, R. W. 1994. Can local incentives alter a metropolitan city's economic development? *Urban Studies* 31(8):1251-1278.

WBCSD (World Business Council for Sustainable Development). 2009. *Greenhouse Gas Mitigation in the Cement Industry.* Available at *http://www.wbcsd.org/DocRoot/XKkAR9Xv28jqGCS6oebc/WBCSDSectoralApproach.pdf.* Accessed February 16, 2010.

Wear, D. N., and B. C. Murray. 2004. Federal timber restrictions, interregional spillovers, and the impact on U.S. softwood markets. *Journal of Environmental Economics and Management* 47(2):307-330.

Weitzman, M. L. 1974. Prices vs. quantities. *The Review of Economic Studies* 41(4):477-491.

West, J. J., A. M. Fiore, L. W. Horowitz, and D. L. Mauzerall. 2006. Global health benefits of mitigating ozone pollution with methane emission controls. *Proceedings of the National Academy of Sciences of the United States of America* 103(11):3988-3993.

Weyant, J. P., F. C. de la Chesnaye, and G. Blanford. 2006. Overview of EMF-21: Multigas mitigation and climate policy. *Energy Journal* (Special Issue):1-32.

Whitfield, S. C., E. A. Rosa, A. Dan, and T. Dietz. 2009. The future of nuclear power: Value orientations and risk perception. *Risk Analysis* 29(3):425-437.

WHO (World Health Organization). 2005. *Indoor Air Pollution and Health.* Available at *http://www.who.int/mediacentre/factsheets/fs292/en/.* Accessed April 13, 2010.

Wigley, T., L. Clarke, J. Edmonds, H. Jacoby, S. Paltsev, H. Pitcher, J. Reilly, R. Richels, M. Sarofim, and S. Smith. 2009. Uncertainties in climate stabilization. *Climatic Change* 97(1-2):85-121.

Wilson, E. J., S. J. Friedmann, and M. F. Pollak. 2007. Research for deployment: Incorporating risk, regulation and liability for carbon capture and sequestration. *Environmental Science and Technology* 41(17).

Wiser, R., M. Bolinger, and G. Barbose. 2007. Using the federal production tax credit to build a durable market for wind power in the United States. *Electricity Journal* 20(9):77-88.

Wolman, H., and D. Spitzley. 1996. The politics of local economic development. *Economic Development Quarterly* 10(2):115-150.

Worrell, E., and G. Biermans. 2005. Move over! Stock turnover, retrofit and industrial energy efficiency. *Energy Policy* 33(7):949-962.

Worrell, E., and C. Galitsky. 2004. *Energy Efficiency Improvement Opportunities for Cement Making: An Energy Star Guide for Energy and Plant Managers.* Berkeley, CA: Lawrence Berkeley National Laboratory, University of California.

Wren, C. 1987. The relative effects of local authority financial assistance policies. *Urban Studies* 24(4):268-278.

WRI (World Resources Institute). 2007. *Slicing the Pie: Sector-based Approaches to International Climate Agreements: Issues and Options.* R. Bradley, K. A. Baumert, B. Childs, T. Herzog, and J. Pershing, eds. Washington, DC: WRI.

WRI. 2009. *Climate Analysis Indicators Tool, Ver 6.* Available from *http://cait.wri.org/.*

WRI. 2010. *Climate Analysis Indicators Tool (CAIT) Version 7.0.* Washington, DC: WRI.

Wu, J. 2000. Slippage effects of the conservation reserve program. *American Journal of Agricultural Economics* 82(4):979-992.

Wu, J., and W. G. Boggess. 1999. The optimal allocation of conservation funds. *Journal of Environmental Economics and Management* 38(3):302-321.

Zimmerman, R. 1993. Social equity and environmental risk. *Risk Analysis* 13(6):649-666.

America's Climate Choices: Membership Lists

COMMITTEE ON AMERICA'S CLIMATE CHOICES

ALBERT CARNESALE (Chair), University of California, Los Angeles
WILLIAM CHAMEIDES (Vice Chair), Duke University, Durham, North Carolina
DONALD F. BOESCH, University of Maryland Center for Environmental Science, Cambridge
MARILYN A. BROWN, Georgia Institute of Technology, Atlanta
JONATHAN CANNON, University of Virginia, Charlottesville
THOMAS DIETZ, Michigan State University, East Lansing
GEORGE C. EADS, Charles River Associates, Washington, D.C.
ROBERT W. FRI, Resources for the Future, Washington, D.C.
JAMES E. GERINGER, Environmental Systems Research Institute, Cheyenne, Wyoming
DENNIS L. HARTMANN, University of Washington, Seattle
CHARLES O. HOLLIDAY, JR., DuPont, Wilmington, Delaware
KATHARINE L. JACOBS,* Arizona Water Institute, Tucson
THOMAS KARL,* National Oceanic and Atmospheric Administration, Asheville, North Carolina
DIANA M. LIVERMAN, University of Arizona, Tuscon and University of Oxford, United Kingdom
PAMELA A. MATSON, Stanford University, California
PETER H. RAVEN, Missouri Botanical Garden, St. Louis
RICHARD SCHMALENSEE, Massachusetts Institute of Technology, Cambridge
PHILIP R. SHARP, Resources for the Future, Washington, D.C.
PEGGY M. SHEPARD, WE ACT for Environmental Justice, New York, New York
ROBERT H. SOCOLOW, Princeton University, New Jersey
SUSAN SOLOMON, National Oceanic and Atmospheric Administration, Boulder, Colorado
BJORN STIGSON, World Business Council for Sustainable Development, Geneva, Switzerland

Asterisks (*) denote members who resigned during the study process

THOMAS J. WILBANKS, Oak Ridge National Laboratory, Tennessee
PETER ZANDAN, Public Strategies, Inc., Austin, Texas

PANEL ON LIMITING THE MAGNITUDE OF FUTURE CLIMATE CHANGE

ROBERT W. FRI (Chair), Resources for the Future, Washington, D.C.
MARILYN A. BROWN (Vice Chair), Georgia Institute of Technology, Atlanta
DOUG ARENT, National Renewable Energy Laboratory, Golden, Colorado
ANN CARLSON, University of California, Los Angeles
MAJORA CARTER, Majora Carter Group, LLC, Bronx, New York
LEON CLARKE, Joint Global Change Research Institute (Pacific Northwest National Laboratory/University of Marland), College Park, Maryland
FRANCISCO DE LA CHESNAYE, Electric Power Research Institute, Washington, D.C.
GEORGE C. EADS, Charles River Associates, Washington, D.C.
GENEVIEVE GIULIANO, University of Southern California, Los Angeles
ANDREW J. HOFFMAN, University of Michigan, Ann Arbor
ROBERT O. KEOHANE, Princeton University, New Jersey
LOREN LUTZENHISER, Portland State University, Oregon
BRUCE MCCARL, Texas A&M University, College Station
MACK MCFARLAND, DuPont, Wilmington, Delaware
MARY D. NICHOLS, California Air Resources Board, Sacramento
EDWARD S. RUBIN, Carnegie Mellon University, Pittsburgh, Pennsylvania
THOMAS H. TIETENBERG, Colby College (retired), Waterville, Maine
JAMES A. TRAINHAM, RTI International, Research Triangle Park, North Carolina

PANEL ON ADAPTING TO THE IMPACTS OF CLIMATE CHANGE

KATHARINE L. JACOBS* (Chair, through January 3, 2010), University of Arizona, Tucson
THOMAS J. WILBANKS (Chair), Oak Ridge National Laboratory, Tennessee
BRUCE P. BAUGHMAN, IEM, Inc., Alabaster, Alabama
ROBERT BEACHY,* Donald Danforth Plant Sciences Center, Saint Louis, Missouri
GEORGES C. BENJAMIN, American Public Health Association, Washington, D.C.
JAMES L. BUIZER, Arizona State University, Tempe
F. STUART CHAPIN III, University of Alaska, Fairbanks
W. PETER CHERRY, Science Applications International Corporation, Ann Arbor, Michigan
BRAXTON DAVIS, South Carolina Department of Health and Environmental Control, Charleston
KRISTIE L. EBI, IPCC Technical Support Unit WGII, Stanford, California

JEREMY HARRIS, Sustainable Cities Institute, Honolulu, Hawaii
ROBERT W. KATES, Independent Scholar, Bangor, Maine
HOWARD C. KUNREUTHER, University of Pennsylvania Wharton School of Business, Philadelphia
LINDA O. MEARNS, National Center for Atmospheric Research, Boulder
PHILIP MOTE, Oregon State University, Corvallis
ANDREW A. ROSENBERG, Conservation International, Arlington, Virginia
HENRY G. SCHWARTZ, JR., Jacobs Civil (retired), Saint Louis, Missouri
JOEL B. SMITH, Stratus Consulting, Inc., Boulder, Colorado
GARY W. YOHE, Wesleyan University, Middletown, Connecticut

PANEL ON ADVANCING THE SCIENCE OF CLIMATE CHANGE

PAMELA A. MATSON (Chair), Stanford University, California
THOMAS DIETZ (Vice Chair), Michigan State University, East Lansing
WALEED ABDALATI, University of Colorado at Boulder
ANTONIO J. BUSALACCHI, JR., University of Maryland, College Park
KEN CALDEIRA, Carnegie Institution of Washington, Stanford, California
ROBERT W. CORELL, H. John Heinz III Center for Science, Economics and the Environment, Washington, D.C.
RUTH S. DEFRIES, Columbia University, New York, New York
INEZ Y. FUNG, University of California, Berkeley
STEVEN GAINES, University of California, Santa Barbara
GEORGE M. HORNBERGER, Vanderbilt University, Nashville, Tennessee
MARIA CARMEN LEMOS, University of Michigan, Ann Arbor
SUSANNE C. MOSER, Susanne Moser Research & Consulting, Santa Cruz, California
RICHARD H. MOSS, Joint Global Change Research Institute (Pacific Northwest National Laboratory/University of Marland), College Park, Maryland
EDWARD A. PARSON, University of Michigan, Ann Arbor
A. R. RAVISHANKARA, National Oceanic and Atmospheric Administration, Boulder, Colorado
RAYMOND W. SCHMITT, Woods Hole Oceanographic Institution, Massachusetts
B. L. TURNER II, Arizona State University, Tempe
WARREN M. WASHINGTON, National Center for Atmospheric Research, Boulder, Colorado
JOHN P. WEYANT, Stanford University, California
DAVID A. WHELAN, The Boeing Company, Seal Beach, California

PANEL ON INFORMING EFFECTIVE DECISIONS AND
ACTIONS RELATED TO CLIMATE CHANGE

DIANA LIVERMAN (Co-chair), University of Arizona, Tucson

PETER RAVEN (Co-chair), Missouri Botanical Garden, Saint Louis

DANIEL BARSTOW, Challenger Center for Space Science Education, Alexandria, Virginia

ROSINA M. BIERBAUM, University of Michigan, Ann Arbor

DANIEL W. BROMLEY, University of Wisconsin-Madison

ANTHONY LEISEROWITZ, Yale University

ROBERT J. LEMPERT, The RAND Corporation, Santa Monica, California

JIM LOPEZ,* King County, Washington

EDWARD L. MILES, University of Washington, Seattle

BERRIEN MOORE III, Climate Central, Princeton, New Jersey

MARK D. NEWTON, Dell, Inc., Round Rock, Texas

VENKATACHALAM RAMASWAMY, National Oceanic and Atmospheric Administration, Princeton, New Jersey

RICHARD RICHELS, Electric Power Research Institute, Inc., Washington, D.C.

DOUGLAS P. SCOTT, Illinois Environmental Protection Agency, Springfield

KATHLEEN J. TIERNEY, University of Colorado at Boulder

CHRIS WALKER, The Carbon Trust LLC, New York, New York

SHARI T. WILSON, Maryland Department of the Environment, Baltimore

Panel on Limiting the Magnitude of Future Climate Change: Statement of Task

The Panel on Limiting the Magnitude of Future Climate Change will describe, analyze, and assess strategies for reducing the net future human influence on climate, including both technology and policy options (sometimes referred to as "mitigation of climate change"). The panel will focus on actions to reduce domestic greenhouse gas emissions and other human drivers of climate change, such as changes in land use, but will also consider the international dimensions of climate stabilization. The panel will not be responsible for evaluating specific proposals to limit or counteract climate change via direct interventions in the climate system (i.e., so-called geoengineering approaches) but may comment on the possible role that such approaches could play in future plans to limit the magnitude of climate change. The panel will also strive to keep abreast of the wide range of proposals currently being advanced by policy makers at a number of levels to limit the future magnitude of climate change, and strive to frame their recommendations in the context of these developments.

The panel will be challenged to produce a report that is broad and authoritative, yet concise and useful to decision makers. The costs, benefits, limitations, trade-offs, and uncertainties associated with different options and strategies should be assessed qualitatively and, to the extent practicable, quantitatively, using the scenarios of future climate change and vulnerability provided by the Climate Change Study Committee. The panel should also provide policy-relevant (but not policy-prescriptive) input to the committee on the following overarching questions:

- What short-term actions can be taken to limit the magnitude of future climate change?
- What promising long-term strategies, investments, and opportunities could be pursued to limit the magnitude of future climate change?
- What are the major scientific and technological advances (e.g., new observations, improved models, research priorities, etc.) needed to limit the magnitude of future climate change?

- What are the major impediments (e.g., practical, institutional, economic, ethical, intergenerational, etc.) to limiting the magnitude of future climate change, and what can be done to overcome these impediments?
- What can be done to limit the magnitude of future climate change at different levels (e.g., local, state, regional, national, and in collaboration with the international community) and in different sectors (e.g., nongovernmental organizations, the business community, the research and academic communities, individuals and households, etc.)?

Panel on Limiting the Magnitude of Future Climate Change: Biographical Sketches

Mr. Robert W. Fri (Chair) is a visiting scholar and senior fellow emeritus at Resources for the Future, a nonprofit organization that studies natural resource and environmental issues. He has served as director of the National Museum of Natural History, president of Resources for the Future, and deputy administrator of both the Environmental Protection Agency and the Energy Research and Development Administration. Fri has been a director of the American Electric Power Company and vice-chair and a director of the Electric Power Research Institute. He is a trustee and vice-chair of Society for Science and the Public, and a member of the National Petroleum Council. He is active with the National Academies, where he is National Associate, vice-chair of the Board on Energy and Environmental Systems, and a member of the Advisory Board of the Marion E. Koshland Science Museum. He has chaired studies for the National Research Council on the health standards for the Yucca Mountain repository, on estimating the benefits of applied research programs at the Department of Energy (DOE), and on evaluating the nuclear energy research program at DOE. Fri received his B.A. in physics from Rice University and his M.B.A. from Harvard University, and he is a member of Phi Beta Kappa and Sigma Xi.

Dr. Marilyn A. Brown is an endowed Professor of Energy Policy in the School of Public Policy at the Georgia Institute of Technology, which she joined in 2006 after a distinguished career at the U.S. Department of Energy's Oak Ridge National Laboratory (ORNL). At ORNL, she held various leadership positions and co-led the report *Scenarios for a Clean Energy Future*, which remains a cornerstone of engineering-economic analysis of low-carbon energy options for the United States. Her research interests encompass the design of energy and climate policies, issues surrounding the commercialization of new technologies, and methods for evaluating sustainable energy programs and policies. Dr. Brown has authored more than 200 publications including a recently published book, *Energy and American Society: Thirteen Myths*, and a forthcoming book, *Climate Change and Energy Security*. Dr. Brown has been an expert witness in hearings before committees of both the U.S. House of Representatives and the U.S. Senate, and she participates on several National Academies boards and committees. Dr. Brown has a Ph.D. in geography from the Ohio State University, and a master's degree in resource planning from the University of Massachusetts and is a Certified Energy Manager.

Dr. Doug Arent is Executive Director of the Joint Institute for Strategic Energy Analysis at the National Renewable Energy Laboratory (NREL). He specializes in strategic planning and financial analysis, clean energy technologies and energy and water issues, and international and governmental policies. In addition to his NREL responsibilities, Arent is an author and expert reviewer for the Intergovernmental Panel on Climate Change (IPCC) Special Report on Renewable Energy, a member of the U.S. Government Review Panel for the IPCC Reports on Climate Change, and a Senior Visiting Fellow at the Center for Strategic and International Studies. Arent is on the Executive Council of the U.S. Association of Energy Economists, is a Member of the Keystone Energy Board, and is on the Advisory Board of E+Co, a public-purpose investment company that supports sustainable development across the globe. He serves on the University of Colorado Chancellor's Committee on Energy, Environment and Sustainability Carbon Neutrality Group and on the Clean and Diversified Energy Advisory Council of the Western Governor's Association. Previously, Arent was Director of the Strategic Energy Analysis Center at NREL and was a management consultant to several clean energy companies. Dr. Arent has a Ph.D. from Princeton University and an M.B.A. from Regis University.

Ms. Ann Carlson is Professor of Law and the inaugural Faculty Director of the Emmett Center on Climate Change and the Environment at the University of California, Los Angeles (UCLA) School of Law. She is also on the faculty of the UCLA Institute of the Environment. Professor Carlson's scholarship in environmental law focuses on climate change law and policy, federalism, and the role social norms play in affecting environmentally cooperative behavior. Her recent work involves analyzing unusual models of environmental federalism, with a focus on the unique role California plays in regulating mobile source emissions, including greenhouse gas emissions, under the Clean Air Act. She has also written on the legal and political obstacles utilities will face in cutting greenhouse gas emissions and on the threat of heat waves and climate change. She is a frequent commentator and speaker on environmental issues, particularly on climate change. Professor Carlson's article "Takings on the Ground" was selected in 2003 by the *Land Use and Environmental Law Review* as one of the top 10 environmental articles of the year. She is co-author (with Daniel Farber and Jody Freeman) of *Environmental Law* (7th Ed.). Professor Carlson teaches Property, Environmental Law and Climate Change Law and Policy and was the recipient of the 2006 Rutter Award for Excellence in Teaching. She served as the law school's academic associate dean from 2004 to 2006. Carlson received her J.D. magna cum laude from Harvard Law School in 1989 and her B.A., magna cum laude, from the University of California at Santa Barbara in 1982.

Ms. Majora Carter, from 2001 to 2008, was the Founder and Executive Director of Sustainable South Bronx (SSBx), a nonprofit environmental justice solutions corporation that designs and implements economically viable and innovative projects that are informed by community needs. She recently moved on from SSBx to form the Majora Carter Group LLC, a green-collar economic consulting firm. Carter led efforts to create the South Bronx Greenway—11 miles of alternative transport, local economic development, low-impact stormwater management, and recreational space—as well as a highly successful effort to create intensive urban forestation, green roofing and walls, and water-permeable open spaces. In 2003, SSBx started the Bronx Environmental Stewardship Training program, one of the nation's first urban green-collar job training and placement systems. Her local and global environmental solutions rest on poverty alleviation through green economic development, and empowering communities to resist the bad environmental decisions that have led to both public health and global atmospheric problems. She is a 2006 MacArthur "genius" Fellow, one of *Essence* Magazine's 25 most influential African Americans for 2007, co-host of the Green on the Sundance Channel, and host of the public radio series *The Promised Land*.

Dr. Leon Clarke is a Senior Research Economist at the Pacific Northwest National Laboratory (PNNL), and he is a staff member of the Joint Global Change Research Institute (JGCRI), a collaboration between PNNL and the University of Maryland at College Park. Dr. Clarke's current research focuses on the role of technology in addressing climate change, international climate policy, scenario analysis, and integrated assessment model development. Dr. Clarke coordinated the U.S. Climate Change Science Program's emissions scenario development process (SAP 2.1a), he was a contributing author on the Working Group III contribution to the IPCC's Fourth Assessment Report, and he coordinated and co-edited the Energy Modeling Forum 22 Transition Scenarios project. He is currently a lead author on the IPCC's *Special Report on Renewable Energy*. Prior to joining PNNL, Dr. Clarke worked for Lawrence Livermore National Laboratory, Pacific Gas & Electric Company, and RCG/Hagler, Bailly, Inc. He was also a research assistant at Stanford's Energy Modeling Forum. Dr. Clarke received B.S. and M.S. degrees in mechanical engineering from U.C. Berkeley and M.S. and Ph.D. degrees in engineering economic systems and operations research from Stanford University.

Mr. Francisco de la Chesnaye is a Senior Economist and Policy Analyst in the Global Climate Change Program at the Electric Power Research Institute (EPRI). He manages EPRI's Regional Modeling project, which is developing a new U.S. energy-economic model to assess the impact of climate and energy policies on the electric power sector, the energy system, and the economy at both regional and national scales. Prior to joining EPRI, Mr. de la Chesnaye was the Chief Climate Economist at the U.S. Environmental Protection Agency (EPA). He was responsible for developing and applying EPA's economic models for domestic and international climate change policy analysis. He led EPA's efforts to produce the agency's first independent economic analysis of a climate policy, the McCain-Lieberman bill of 2007. Mr. de la Chesnaye was a Lead Author for the IPCC's Fourth Assessment Report and was a co-editor of *Human-Induced Climate Change: An Interdisciplinary Assessment* (Cambridge University Press, 2007). Mr. de la Chesnaye is currently pursuing a Ph.D. at the University of Maryland. He earned an M.S. from Johns Hopkins in 2002, an M.A. from American University in 1997, and a B.S. from Norwich University, The Military College of Vermont, in 1988.

Dr. George C. Eads is a Senior Consultant of Charles River Associates (CRA). Prior to joining CRA in 1995, he held several positions at General Motors (GM) Corporation, including Vice President and Chief Economist; Vice President, Worldwide Economic and Market Analysis Staff; and Vice President, Product Planning and Economics Staff. Before joining GM, Dr. Eads was Dean of the School of Public Affairs at the University of Maryland, College Park, where he also was a professor. Before that, he served as a Member of President Carter's Council of Economic Advisors. He has been involved in numerous projects concerning transport and energy. In 1994 and 1995, he was a member of President Clinton's policy dialogue on reducing greenhouse gas emissions from personal motor vehicles. He co-authored the World Energy Council's 1998 Report, *Global Transport and Energy Development—The Scope for Change.* Over the past 4 years, Dr. Eads devoted most of his time to the World Business Council for Sustainable Development's Sustainable Mobility Project, a project funded and carried out by 12 leading international automotive and energy companies. Dr. Eads is a member of the Presidents' Circle at the National Academies. He is an at-large Director of the National Bureau of Economic Research. He received a Ph.D. degree in economics from Yale University. He is currently participating in a Transportation Research Board (TRB) study *Potential Greenhouse Gas Reductions from Transportation* and recently completed service on the TRB study *Climate Change and U.S. Transportation.*

Dr. Genevieve Giuliano is Professor and Senior Associate Dean of Research and Technology in the School of Policy, Planning, and Development, University of Southern California (USC), and Director of the METRANS joint USC and California State University Long Beach Transportation Center. She also holds courtesy appointments in civil engineering and geography. Professor Giuliano's research focus areas include relationships between land use and transportation, transportation policy analysis, and information technology applications in transportation. She has published over

130 papers and has presented her research at numerous conferences both within the United States and abroad. She serves on the editorial boards of *Urban Studies* and the *Journal of Transport Policy*. She is a past member and Chair of the Executive Committee of the Transportation Research Board (TRB). She was named a National Associate of the National Academy of Sciences in 2003, received the TRB William Carey Award for Distinguished Service in 2006, and was awarded the Deen Lectureship in 2007. She has participated in several National Research Council policy studies; currently she is on the Committee for Global Climate Change and Transportation. She was recently appointed Chair of the California Research and Technology Advisory Panel, which will advise both Caltrans and the Department of Business, Housing and Transportation on the implementation of the Growth Management Plan.

Dr. Andrew Hoffman is the Holcim (U.S.) Professor of Sustainable Enterprise; Professor of Natural Resources; Associate Professor of Management & Organizations; and Associate-Director of the Erb Institute for Global Sustainable Enterprise, at the University of Michigan. He studies organizational culture, values, and behavior, with a particular emphasis on market drivers and corporate strategies for addressing climate change. Previously, he was Associate Professor of Organizational Behavior at the Boston University School of Management and Senior Fellow at the Meridian Institute; he also held positions at the Amoco Oil Company, T&T Construction and Design, Metcalf & Eddy, and the Environmental Protection Agency (Region 1). Dr. Hoffman has written numerous books and articles about corporate strategies for addressing climate change and has organized and moderated conferences that brought together senior executives from business, government, and the environmental community to discuss the scientific, strategic, and policy implications of controls on greenhouse gas emissions. He has a Ph.D. from MIT's Alfred P. Sloan School of Management and Department of Civil and Environmental Engineering (interdepartmental degree) and a B.S. in chemical engineering from the University of Massachusetts at Amherst.

Dr. Robert O. Keohane is Professor of International Affairs, Princeton University. He is the author of *After Hegemony: Cooperation and Discord in the World Political Economy* (1984) and *Power and Governance in a Partially Globalized World* (2002). He is coauthor of *Power and Interdependence* (1977 and subsequent editions) and of *Designing Social Inquiry* (1994). He has served as the editor of the journal *International Organization* and as president of the International Studies Association and the American Political Science Association. He is a member of the American Academy of Arts and Sciences, the American Philosophical Society, and the National Academy of Sciences. He has received honorary degrees from the University of Aarhus, Denmark, and Sciences Po in Paris, and is a fellow of the American Academy of Political and Social Science. His recent work has involved political analyses of "post-Kyoto" climate change architecture. He earned his B.A. from Shimer College in 1961 and his Ph.D. from Harvard in 1966.

Dr. Loren Lutzenhiser is Professor of Urban Studies and Planning. He has a Ph.D. in sociology. Dr. Lutzenhiser's teaching interests include environmental policy and practice, energy behavior and climate, technological change, urban environmental sustainability, and social research methods. His research focuses on the environmental impacts of sociotechnical systems, particularly how urban energy/resource use is linked to global environmental change. Particular studies have considered variations across households in energy consumption practices, how energy-using goods are procured by government agencies, how commercial real estate markets work to develop both poorly performing and environmentally exceptional buildings, and how the "greening" of business may be influenced by local sustainability movements and business actors. He recently completed a major study for the California Energy Commission reporting on the behavior of households, businesses, and governments in the aftermath of that state's 2001 electricity deregulation crisis. He is currently exploring the relationships between household natural gas, electricity, gasoline, and water usage.

Dr. Bruce McCarl is Distinguished Professor of Agricultural Economics at Texas A&M University. He has a Ph.D. in management science from Pennsylvania State University. His research efforts involve policy analysis in climate change, climate change mitigation, and El Niño/Southern Oscillation analysis and water resource issues, as well as the proper application of quantitative methods to such analyses. Dr. McCarl began work on the agricultural and forestry effects of climate change in the 1980s, including the role agriculture and forestry could play in mitigating climate change through sequestration, GHG emissions offsets, or emissions reduction. Dr. McCarl has also been addressing agriculture and bioenergy since the late 1970s. He developed the first sector-wide economic appraisal of bioenergy prospects from agriculture and led OTA analyses of corn and cellulosic ethanol, well in advance of today's activity. Recently McCarl has worked on greenhouse gas implications of producing biodiesel, ethanol, cellulosic ethanol, and biofeedstock-fueled electric power. McCarl's research has also encompassed water resources, including groundwater management, irrigation concerns in the agricultural sector, and climate change analyses. He served as a lead author for the IPCC, Agricultural Mitigation, Working Group III.

Dr. Mack McFarland is an Environmental Fellow for DuPont Fluoroproducts. He received a Ph.D. in chemical physics from the University of Colorado, was a postdoctoral fellow at York University and then a research scientist at the NOAA Aeronomy Laboratory. Mack planned, conducted, and interpreted field experiments designed to probe the cycles that control atmospheric ozone concentrations. These studies included measurements of gases and processes important to the global climate change issue. In late 1983, Mack joined the DuPont Company. His primary responsibilities have been in the areas of coordination of research programs and assessment and interpretation of scientific information on stratospheric ozone depletion and global climate change. During 1995 and 1996, Mack was on loan to the Atmosphere Unit of the United Nations Environment Programme and in 1997 he was on loan to the IPCC Working Group II Technical Support Unit. The value of his contributions to DuPont has been recognized through a C&P Flagship Award, Environmental Respect Awards, and Environmental Excellence Awards. In 1999, Mack was awarded an individual Climate Protection Award by the U.S. Environmental Protection Agency for his contributions in providing understandable, reliable information to decision makers.

Ms. Mary D. Nichols was appointed by Governor Arnold Schwarzenegger as Chairman of the California Air Resources Board in July 2007. She returned to the Air Board 30 years after serving as the Chairman under Governor Jerry Brown from 1978 to 1983. Nichols has devoted her entire career in public and private, not-for-profit service to advocating for the environment and public health. In addition to her work at the Air Board, she has held a number of positions, including assistant administrator for the EPA's Air and Radiation program under the Clinton Administration, Secretary for California's Resources Agency from 1999 to 2003, and Director of the University of California, Los Angeles, Institute of the Environment. As one of California's first environmental lawyers, she initiated precedent-setting test cases under the Federal Clean Air Act and California air quality laws while practicing as a staff attorney for the Center for Law in the Public Interest. Nichols holds a J.D. degree from Yale Law School and a B.A. degree from Cornell University.

Dr. Edward S. Rubin is a professor in the Departments of Engineering and Public Policy, and Mechanical Engineering, at Carnegie Mellon University. He holds a chair as the Alumni Professor of Environmental Engineering and Science and was founding director of the university's Center for Energy and Environmental Studies and the Environmental Institute. His teaching and research are in the areas of energy utilization, environmental control, technology innovation, and technology-policy interactions, with a particular focus on issues related to coal utilization, carbon sequestration, and global climate change. He is the author of over 200 technical publications and a textbook, *Introduction to Engineering and the Environment*. He is a fellow member of ASME, a past chairman of its Environmental Control Division, and recipient of the Air & Waste Management Association Lyman A. Ripperton Award for distinguished achievements as an educator and the Distinguished Professor of Engineering Award from Carnegie Mellon University. He has served as an advisor to government agencies including the U.S. Department of Energy and the U.S. Environmental Protection Agency, and on various committees of the National Academies, including the Board on Energy and Environmental Systems, the 1992 study, Policy Implications of Global Warming, and recent studies of coal research and development needs and the potential for hydrogen-powered vehicles. He was also a coordinating lead author of the IPCC Special Report on Carbon Dioxide Capture and Storage. Dr. Rubin received his bachelor's degree in mechanical engineering from the City College of New York and his master's and Ph.D. degrees from Stanford University.

Dr. Thomas H. Tietenberg recently retired as the Mitchell Family Professor of Economics at Colby College. Specializing in environmental and natural resource economics, his areas of expertise include emissions trading, climate change policy, and economic incentives for pollution control. He is the author or editor of 11 books (including *Environmental and Natural Resource Economics*, one of the most widely used textbooks in the field), as well as over 100 articles and essays on environmental and natural resource economics. Former president and current fellow of the Association of Environmental and Resource Economists, he has consulted on environmental policy with a number of international organizations as well as several state and foreign governments. Dr. Tietenberg is currently serving as one of three appointed trustees for the Energy and Carbon Savings Trust, an organization that receives all Maine revenues from the sale of carbon allowances in the Northeast's Regional Greenhouse Gas Initiative and uses them to promote energy efficiency in the state.

Dr. James A. Trainham is Vice President, Strategic Energy Initiatives, for RTI International, with a joint appointment to North Carolina State University. His focus is the research and development of solar fuels. Most recently (2008–2010), Trainham led the successful pilot demonstration of the first solar thermal biomass gasifier as Sundrop Fuels, Inc. Senior Vice President Engineering with responsibility for R&D, engineering design, scale-up, and commercialization of Sundrop Fuels' unique solar gasification technology. Previously, he served as Vice-President of Science and Technology for PPG Industries, one of the world's leading coatings and materials manufacturing companies. He also served as global technology director of the DuPont Company and was responsible for R&D for new products, new processes, and fundamental and end-use research carried out in five laboratories, and intellectual property management. A member of management since 1983, Dr. Trainham remains a technology innovator with over 40 patents and publications. He was elected to the National Academy of Engineering in 1997 and was recently honored by the American Institute of Chemical Engineers as one of the "One Hundred Chemical Engineers of the Modern Era" for his leadership in sustainability. Dr. Trainham received his B.S. and Ph.D. degrees in chemical engineering from the University of California, Berkeley, and an M.S. degree in chemical engineering from the University of Wisconsin, Madison. He recently chaired the NRC Committee on Grand Challenges for Sustainability in the Chemical Industry.

Acronyms, Energy Units, and Chemical Formulas

ACRONYMS

AEF	America's Energy Future
ARRA	American Recovery and Reinvestment Act
BEA	Bureau of Economic Analysis
BEST	Bronx Environmental Stewardship Training
BTAs	Border Tax Adjustments
CAA	Clean Air Act
CAFE	Corporate Average Fuel Economy
CBO	Congressional Budget Office
CCCSTI	Committee on Climate Change Science and Technology Integration
CCS	Carbon capture and storage
CDM	Clean Development Mechanism
CER	Certified emission reduction
DOD	U.S. Department of Defense
DOE	U.S. Department of Energy
DOI	U.S. Department of Interior
DOS	U.S. Department of State
EIA	Energy Information Administration
EISA	Energy Independence and Security Act
EMF	Energy Modeling Forum
EPA	Environmental Protection Agency
EPRI	Electric Power Research Institute
EU ETS	European Union Emissions Trading Scheme
GAO	Government Accountability Office
GDP	Gross domestic product
GHG	Greenhouse gas
GWP	Global warming potential
IEA	International Energy Agency
IPCC	Intergovernmental Panel on Climate Change
LCFS	Low Carbon Fuel Standards

LED   Light-emitting diode
LUCF   Land-use change and forestry
NAAQS   National Ambient Air Quality Standards
NRC   National Research Council
NREL   National Renewable Energy Laboratory
NSF   National Science Foundation
OECD   Organisation for Economic Co-operation and Development
R&D   Research and development
REDD   Reducing emissions from deforestation in developing nations
RFS   Renewable Fuels Standard
RGGI   Regional Greenhouse Gas Initiative
RPS   Renewable Portfolio Standard
S&E   Science and engineering
TRB   Transportation Research Board
UNEP   United Nations Environmental Program
UNFCCC   United Nations Framework Convention on Climate Change
USDA   U.S. Department of Agriculture
VMT   Vehicle miles traveled
WCI   Western Climate Initiative
WTO   World Trade Organization

CHEMICAL COMPOUNDS

CFCs   Chlorofluorocarbons
CH_4 Methane
CO_2 Carbon dioxide
HFC   Hydrofluorocarbons
NMHCs   Nonmethane hydrocarbons
NO_x Nitrogen oxides
O_3 Tropospheric ozone
PFCs   Perfluorocarbons
PM   Particulate matter
N_2O Nitrous oxide
SF_6 Sulfur hexafluoride
SO_2 Sulfur dioxide
VOC   Volatile organic compounds

UNITS USED (FOR ENERGY, POWER, MATTER)

J: joule. The energy of one watt of power flowing for one second
GJ: gigajoule, 10^9 joules
EJ : exajoule, 10^{18} joules
BTU: British thermal unit. The energy to raise the temperature of one pound of water one degree Fahrenheit.

quad: a quadrillion (a million-billion, or 10^{15}) BTUs, equal to 1.055×10^{18} joules (1.055 EJ). A unit commonly used in discussing global and national energy budgets

W: watt, a unit of electric power (= 1 J of energy per second)
kW: kilowatt, a thousand (10^3) Watts
GW: gigawatt, a billion (10^9) Watts
TW: terawatt, a trillion (10^{12}) Watts
kWh: kilowatt-hour, or the amount of energy when one kW is used for one hour. Equivalent to about 3,400 BTU or 3,600,000 Joules

Metric ton (i.e., of CO_2) is one thousand kilograms, or about 2,200 pounds
MMT: million metric tons of CO_2
Mt: megaton, a million (10^6) metric tons
Gt: gigaton, a billion (10^9) metric tons
Tg: teragram, a billion (10^9) kilograms, or one million metric tons

ppm: parts per million, a measure of atmospheric concentration of some greenhouse gases

Barrel (oil): ~42 gallons. The United States uses about 20 million barrels of oil each day.